Applied Codeology

Navigating the NEC 2011

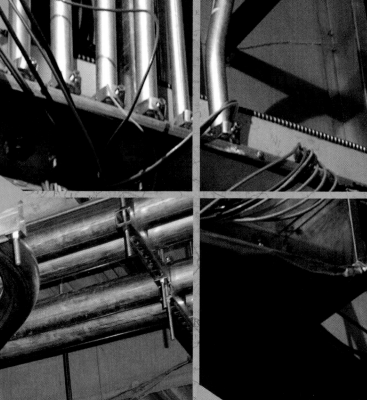

NATIONAL JOINT APPRENTICESHIP
AND TRAINING COMMITTEE

Contents

Contents

Contents

Features

Headers and **Subheaders** organize information within the chapter.

Figures, including photographs and artwork, clearly illustrate concepts from the text.

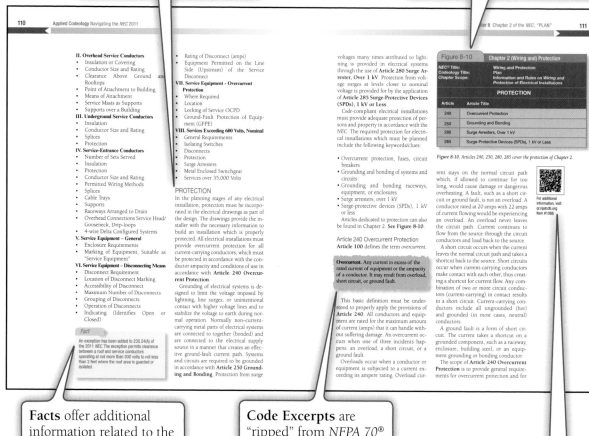

Facts offer additional information related to the *Codeology* method.

Code Excerpts are "ripped" from *NFPA 70*® or other sources.

For additional information related to QR Codes, visit qr.njatcdb.org Item #1079

Quick Response Codes (QR Codes) create a link between the textbook and the Internet. They can be scanned using Smartphone applications to obtain additional information online. (To access the information without using a Smartphone, visit qr.njatc.org and enter the referenced Item #.)

Features

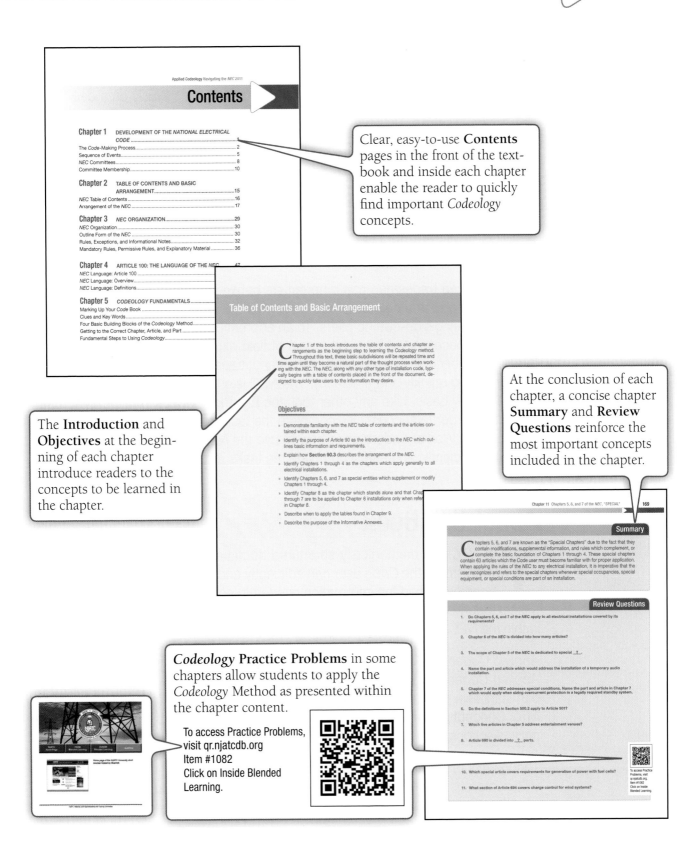

Clear, easy-to-use **Contents** pages in the front of the textbook and inside each chapter enable the reader to quickly find important *Codeology* concepts.

The **Introduction** and **Objectives** at the beginning of each chapter introduce readers to the concepts to be learned in the chapter.

At the conclusion of each chapter, a concise chapter **Summary** and **Review Questions** reinforce the most important concepts included in the chapter.

Codeology **Practice Problems** in some chapters allow students to apply the *Codeology* Method as presented within the chapter content.

To access Practice Problems, visit qr.njatcdb.org
Item #1082
Click on Inside Blended Learning.

Introduction

Apprentices, Journeymen Wiremen, Electrical Engineers, Electrical Maintenance Workers, Electrical Inspectors, and numerous other Electrical Workers will use the NFPA 70: *National Electrical Code® (NEC)* on a daily basis. Their livelihood depends upon their ability to properly design, install, inspect, and maintain electrical systems in accordance with the *NEC*.

The *NEC* can be overwhelming to new apprentices in the electrical industry. In fact, some Electrical Workers complete an apprenticeship or an electrical engineering degree without confidence in their ability to quickly find references within the *NEC*. This can be very frustrating since they must use the *NEC* on a daily basis to complete their work. Add to this the fact that the *NEC* changes every three years and this frustration continues to grow.

An experienced tradesman can differentiate between members of different trades just by looking at the tools they carry. Electrical Workers use hand tools, the most identifiable being side-cutting pliers. To an apprentice, a pair of side-cutters seems clumsy, but necessary. After a few years of field experience, those same side-cutters seem to become an extension of the hand, capable of many tasks; all of which are performed quickly and efficiently. The *NEC* is a "tool" which Electrical Workers use on a daily basis. Like any other tool, the Electrical Worker must learn to use it properly in order to successfully complete the task required. When and only when the Electrical Worker understands how to use the *NEC*, he/she will be able to successfully, efficiently, and safely complete the tasks required of them.

About this Book

The *NEC* is a methodically structured document. Once this structure is fully understood, the *Code* user can efficiently and confidently maneuver through it and easily find the information needed to complete a task. The objective of this textbook is to build confidence in the ability to quickly find pertinent information within the *NEC*.

Applied Codeology will describe the fundamental structure of the *NEC*. In addition, *Applied Codeology* will introduce important concepts which the Electrical Worker can start using immediately to improve the skills needed to find information in the *NEC*. Concepts such as Plan, Build, Use and the importance of highlighting articles, parts, sections, and subdivisions are explained in detail.

Though this textbook is not intended to be "article" or "topic" specific, it does provide numerous practical examples intended to promote the skills necessary to find specific information within the *NEC*.

Acknowledgements

Acknowledgements

Gary Beckstrand
CDI Torque Products
Cooper Bussman
Eric S. Davis
Ideal Industries Inc.
Legrand
National Fire Protection Association
    (NFPA)
Alan Shepard
Thomas & Betts

QR Codes

ARCAT
CDI Torque Products
Cooper Bussman
Erico International
Hubbell Power Systems
Ideal Industries Inc.
Legrand
Lincoln Electric
National Fire Protection Association
    (NFPA)
Rockwell Automation
Sunwize Technologies Inc.
Thomas & Betts

This material is continually reviewed and evaluated by Training Directors who are also members of the NJATC Inside Education Committee. The invaluable input provided by these individuals allows for the development of instructional material that is of the absolute highest quality. At the time of this printing, the Inside Education Committee was composed of the following members:

Kathleen Barber, San Carlos, CA
Byron Benton, San Leandro, CA
John Biondi, Vineland, NJ
Richard M. Brooks, Dayton, OH
Eric S. Davis, Warren, OH
Lawrence Hidalgo, Lansing, MI
Gregory A. Hojdila, Beaver, PA
Kenneth Jania, Merrillville, IN

Tony Lewis, Tacoma, WA
Dave McCraw, Tampa, FL
David Milazzo, Paramus, NJ
Tom Minder, Fairbanks, AK
Tony Naylor, Wichita, KS
Janet Skipper, Winter Park, FL
Jim Sullivan, Winter Park, FL
Dennis Williamson, Kennewick, WA

Development of the *National Electrical Code*

The *National Electrical Code*® (*NEC*) is revised every three years and made available to the *Code* user. Throughout the three years prior to a new release, Technical Committee members and NFPA staff collaborate extensively with task groups through many stages to address existing issues identified in the previous cycle and to address new submittals. These task groups work on technical issues and usability matters to formulate proposals for the next *NEC* cycle. The *NEC* is a dynamic document, a true work in progress which is shaped, molded, and improved by all members of the public who take part in the process.

Objectives

» Outline the structure of the NFPA and *NEC* governing bodies and the meaning of a consensus standard.

» Explain the proposal, comment, and amending motions process in the three-year revision cycle of the *NEC*.

» Identify the makeup of different classifications of a *Code*-Making Panel or Technical Committee.

» Explain how you can take part in the *NEC* revision process.

Chapter 1

Table of Contents

THE *CODE*-MAKING PROCESS

This text, *Applied Codeology*, is an introduction to understanding the NFPA 70 *National Electrical Code*® (NEC). The National Fire Protection Agency (NFPA) was founded in 1896 in Boston by a small group of men who saw the need to set standards for protection of the public against fire hazards. The name has been altered throughout the years, but the initial goal still remains the same—protection of the public. Initially, the organization developed recommended practices for fire suppression using sprinklered water and addressed design and installation of early distribution of electricity, but over the past 100 years the NFPA has become the *Code*-making organization for Fire Protection codes which most U.S. municipalities adopt.

Q. How are the Codes derived and designed?

A. The NFPA Code-making process is not exclusively achieved solely within the organization. In fact, this is quite the contrary. The NFPA codes are completely developed by volunteer users of the code. All the various NFPA codes have an associated Code-Making Panel (CMP) which is closely associated with that particular code area. Members assigned to each CMP work collaboratively to research, define, and submit the code language for public approval. As you will learn in this chapter, members of the electrical industry in the U.S. completely drive the language which makes up the National Electrical Code.

Structure of the NFPA

The National Fire Protection Association (NFPA) derives minimum codes and standards for fire prevention and other life-safety codes and standards for the United States. The NFPA maintains over 300 codes and standards. The NFPA

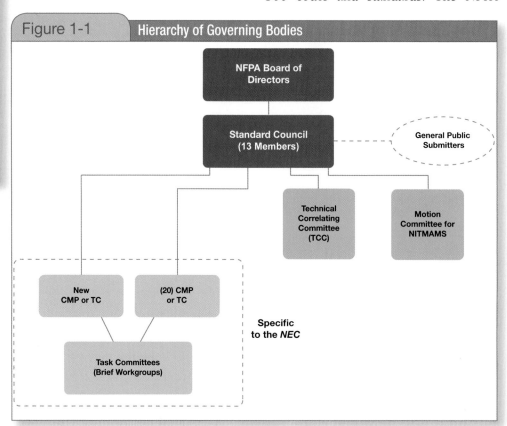

Figure 1-1   Hierarchy of Governing Bodies

Figure 1-1. The NFPA Representative Structure is the same throughout all the Standards and Codes.

Board of Directors is the top governing body over the activities of all 300+ codes and standards. The Board of Directors appoints a Standard Council of 13 members which oversees and administers the rules and regulations. The Standard Council appoints all *Code*-Making Panels (CMPs), Technical Committees (TC), Technical Correlating Committees (TCC), and Motion Committees. In addition, the Standard Council derives new CMPs and TCs if new technologies or situations warrant them. Both the NFPA and *NEC* contain a basic hierarchy of governing bodies. The various governing bodies' responsibilities are discussed later in the chapter. **See Figure 1-1.**

Among the 300+ codes and standards are several electrical related chapters such as NFPA 70 *National Electrical Code*, NFPA 70E *Standard for Electrical Safety in the Workplace*, and NFPA 72 *National Fire Alarm and Signaling Code*. **Figure 1-2** is a partial list of the NFPA codes which the electrical worker may be required to review during the layout or installation stages of a project. Keep in mind that all 300+ codes and committees are revised every three years, but are not on the same three-year schedule. For instance, the NFPA 70 (*NEC*) schedule is 2011, 2014, etc., while the NFPA 72 schedule is 2010, 2013, etc. **See Figure 1-2.**

NEC Process

The *National Electrical Code* (NFPA 70) is developed through a consensus standards development process approved by the American National Standards Institute (ANSI). This process includes input from all interested parties (public) throughout the proposal, comment, and Amending Motions stages. Technical Committees (TCs), also known as *Code*-Making Panels (CMPs), are formed to achieve consensus on all proposed changes or revisions to the *NEC*. These volunteer committees, with members representing various classifications, are balanced to ensure all viewpoints of interest groups have a

voice in the deliberations of issues brought before the *NEC*.

Throughout the revision cycle, the process produces a Report on Proposals (ROP), a Report on Comments (ROC), and Certified Amended Motions for the Association Technical Meeting (NFPA National Meeting). The Technical Committees work diligently to provide the *NEC* user with code text which is practical, easy to read and understand, and enforceable. However, as code changes are applied, users may disagree on the implementation of some of the revised requirements. The intent of the changes can be found in the substantiations, CMP statements, and CMP member comments in both the ROP and ROC stages. All persons submitting a proposal will receive a copy of the ROP in either paper, compact disc format, or on-line. Likewise, those submitting a comment to a proposal will receive a copy of the ROC in those same formats. These documents are an extremely valuable tool for the user of the next edition of the *NEC*. In the back of each edition of the *NEC* there is a "Proposal Form." Anyone can complete this form and submit a proposed change to the next edition of the *NEC*. Proposals may also be submitted on-line at the National Fire Protection Association (NFPA) website, www.nfpa.org.

Tech Fact

Anyone can submit a *Code* proposal by filling out the Proposal Form in the back of the *Code* book and submitting it online.

FORM FOR PROPOSAL FOR 2014 NATIONAL ELECTRICAL CODE®

Reprinted with permission from NFPA 70-2011, *National Electrical Code®*, Copyright© 2010, National Fire Protection Association, Quincy, MA 02169. This reprinted material is not the complete and official position of the NFPA on the referenced subject, which is represented only by the standard in its entirety.

Figure 1-2	Partial (Electrical) Listing of NFPA Codes and Standards
NFPA 1	Fire Code
NFPA 3	Standard on Commissioning and Integrated Testing of Fire Protection and Life Safety Systems
NFPA 20	Standard for the Installation of Stationary Pumps for Fire Protection
NFPA 30A	Code for Motor Fuel Dispensing Facilities and Repair Garages
NFPA 70	National Electrical Code®
NFPA 70A	National Electrical Code® Requirements for One- and Two-Family Dwellings
NFPA 70B	Recommended Practice for Electrical Equipment Maintenance
NFPA 70E	Standard for Electrical Safety in the Workplace®
NFPA 72	National Fire Alarm and Signaling Code
NFPA 73	Electrical Inspection Code for Existing Dwellings
NFPA 75	Standard for the Protection of Information Technology Equipment
NFPA 76	Standard for the Fire Protection of Telecommunications Facilities
NFPA 79	Electrical Standard for Industrial Machinery
NFPA 92	Standard for Smoke Management Systems
NFPA 96	Standard for Ventilation Control and Fire Protection of Commercial Cooking Operations
NFPA 99	Standard for Health Care Facilities
NFPA 101	Life Safety Code®
NFPA 110	Standard for Emergency and Standby Power Systems
NFPA 111	Standard on Stored Electrical Energy Emergency and Standby Power Systems
NFPA 115	Standard for Laser Fire Protection
NFPA 170	Standard for Fire Safety and Emergency Symbols
NFPA 220	Standard on Types of Building Construction
NFPA 262	Standard Method of Test for Flame Travel and Smoke of Wires and Cables for Use in Air-Handling Spaces
NFPA 450	Guide for Emergency Medical Services and Systems
NFPA 496	Standard for Purged and Pressurized Enclosures for Electrical Equipment
NFPA 654	Standard for the Prevention of Fire and Dust Explosions from the Manufacturing, Processing, and Handling of Combustible Particulate Solids
NFPA 730	Guide for Premises Security
NFPA 731	Standard for the Installation of Electronic Premises Security Systems
NFPA 780	Standard for the Installation of Lightning Protection Systems
NFPA 791	Recommended Practice and Procedures for Unlabeled Electrical Equipment Evaluation
NFPA 820	Standard for Fire Protection in Wastewater Treatment and Collection Facilities
NFPA 850	Recommended Practice for Fire Protection for Electric Generating Plants and High Voltage Direct Current Converter Stations
NFPA 853	Standard for the Installation of Stationary Fuel Cell Power Systems
NFPA 900	Building Energy Code
NFPA 5000	Building Construction and Safety Code®

For additional
information, visit
qr.njatcdb.org
Item #1060

Figure 1-2. Although NFPA 70 is most familiar to the electrician, there are many other NFPA codes which apply to electrical installations.

SEQUENCE OF EVENTS

There are several steps in the code-making process which lead to the issuance of an NFPA committee document. The following explains the consensus standards development process. Actual deadlines for each of these steps are listed on the NFPA website www.nfpa.org and in the back of the *NEC*. The Codes and Standards-Making Process will be discussed in great detail. **See Figure 1-3.**

Call for Proposals

The Call for Proposals starts the revision cycle. A public notice requesting interested parties to submit specific written proposals is published in the NFPA News, the U.S. Federal Register, the American National Standards Institute's Standards Action, the NFPA's web site, and other publications. Proposals are submitted before the deadline given in the back of the *NEC*, then organized by NFPA staff and the Standard Council to be forwarded to members of the nineteen *NEC* CMPs (*Code*-Making Panels).

Report on Proposals (ROP)

The CMPs meet at the location and dates stated in the back of the *NEC* to act on all proposals. These meetings are open to all interested persons who want to observe the panel proceedings. Actions at the meeting require only a simple majority of votes.

The actions taken by the CMP in the proposal stage are as follows:

- *Accept.* The panel accepts the proposal as written. No panel statement is required.
- *Reject.* The panel rejects the proposal. A panel statement explaining why is required.
- *Accept in Principle.* The panel accepts the proposal in principle but changes the wording or takes other action to achieve the submitter's intention. A panel statement explaining why is required.
- *Accept in Part.* The panel accepts only part of the proposal and rejects the remainder. A panel statement explaining which part was accepted and why the remainder was rejected is required.

For additional information, visit qr.njatcdb.org Item #1061

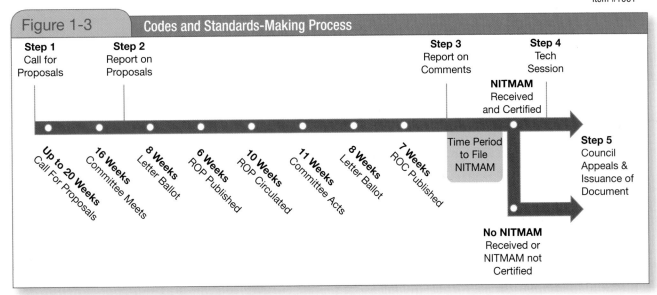

Figure 1-3   Codes and Standards-Making Process

Step 1 Call for Proposals

Step 2 Report on Proposals

Step 3 Report on Comments

Step 4 Tech Session

NITMAM Received and Certified

Up to 20 Weeks Call For Proposals

16 Weeks Committee Meets

8 Weeks Letter Ballot

6 Weeks ROP Published

10 Weeks ROP Circulated

11 Weeks Committee Acts

8 Weeks Letter Ballot

7 Weeks ROC Published

Time Period to File NITMAM

Step 5 Council Appeals & Issuance of Document

No NITMAM Received or NITMAM not Certified

Figure 1-3. There are many steps to code making during the three-year Code cycle.

- *Accept in Principle in Part.* The panel accepts part of the proposal in principle but rejects the remainder. A panel statement is required.

After the CMP meetings have closed, the written ballots are sent to all panel members. A two-thirds majority on the written ballot is required for the panel action to be upheld. The ROP is then printed and distributed free of charge to all proposal submitters and specific CMP.

Report on Comment (ROC)

Comments on the proposal actions must be submitted before the deadline stated in the back of the *NEC*. The comments are then organized by NFPA staff and forwarded to members of the nineteen *Code-*

Making Panels, who thereafter meet at the location and on the dates stated in the back of the *NEC* to act on all comments. These meetings are open to all interested persons who want to observe the panel proceedings. Actions at the meeting require only a simple majority of votes. The actions which may be taken by the CMPs in the comment stage are the same as those in the proposal stage, with one additional action permitted: where a comment introduces new material which has not had public review or would require more time or work to process than allowed, an action of "Hold" is permitted and a panel statement explaining the action is required. A comment which is put on hold returns in the next

Figure 1-4	2014 *NEC Code*-Making Schedule	
PROCESS STAGE	**PROCESS STEP**	**DATES FOR TC**
Preliminary	Notification of Intent to Enter Cycle	7/8/11
Report on Proposals (ROP)	*NEC* Closing Date for Proposals	11/4/11
	NEC Code-Making Panel Meetings (ROP)	1/9-21/12
	Mail *NEC* Ballots to CMPs	1/27/12
	Receipt of Initial *NEC* Ballots	2/24/12
	NEC Correlating Committee Meeting	4/23-27/12
	NEC ROP to Mailing House	7/13/12
Report on Comments (ROC)	*NEC* Closing Date for Comments	10/17/12
	NEC Code-Making Panel Meetings (ROC)	11/28-12/8/12
	Mail *NEC* Ballots to CMPs	12/14/12
	Receipt of *NEC* (TC) Ballots by Staff Liaison	1/11/13
	NEC Correlating Committee Meeting	2/18–22/13
	NEC ROC to Mailing House	3/22/13
Tech Session Preparation & Issuance of Consent Documents	Notice of Intent to Make a Motion (NITMAM) *NEC*	5/3/13
	Posting of Certified *NEC* NITMAMs	5/17/13
Technical Session	Association Meeting for Documents with Certified Amending Motions	6/2-6/13
Appeals & Issuance of Documents w/CAMs	Appeal Closing Date for Documents with Certified Amending Motions	7/3/13
	Council Issuance for Documents with Certified Amending Motions	8/1/13

Figure 1-4. Nearly two years is required to complete all the steps to revise the NEC.

cycle as a proposal. After the CMP meetings have closed, the written ballots are sent to all panel members. A two-thirds majority on the written ballot is required for the panel action to be upheld. The Technical Correlating Committee reviews all of the CMP results to ensure that there are no conflicting actions with other related committees. The ROC is then printed and distributed. **See Figure 1-4.**

NFPA Annual Meeting and Amending Motions

The full *NEC* technical committee report is presented to the NFPA membership for approval at the NFPA Annual Meeting. Since the *NEC* revision is made available to the public for revision years 2011, 2014, etc., the Annual Meetings which include the *NEC* amending motions and approvals are held during the midyears of 2010, 2013, etc. Motions may be made

to amend or reverse the actions taken by the *Code*-Making Panels at the Annual Meeting. Before making a motion at the Annual Association Technical Meeting, the intended maker of the motion must file a Notice of Intent to Make a Motion (NITMAM) 30 days in advance. The cycle schedule will list the deadline date to submit a NITMAM before the Annual Meeting. The NITMAM will be received and approved by the Motions Committee and listed on the approved motion listing as Certified Amending Motions (CAMs) for the Annual Meeting. At the Annual Meeting, the submitter (or representative) of the NITMAM must notify the NFPA prior to one hour of the meeting start time that they will be making the motion. Debate and voting by NFPA members will be allowed at the Annual meeting on all CAMs and will carry with a majority vote. **See Figure 1-5.**

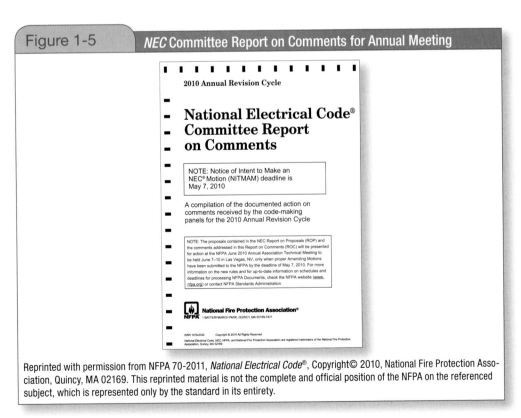

| Figure 1-5 | *NEC* Committee Report on Comments for Annual Meeting |

2010 Annual Revision Cycle

**National Electrical Code®
Committee Report
on Comments**

NOTE: Notice of Intent to Make an
NEC® Motion (NITMAM) deadline is
May 7, 2010

A compilation of the documented action on
comments received by the code-making
panels for the 2010 Annual Revision Cycle

NOTE: The proposals contained in the NEC Report on Proposals (ROP) and the comments addressed in this Report on Comments (ROC) will be presented for action at the NFPA June 2010 Annual Association Technical Meeting to be held June 7–10 in Las Vegas, NV, only when proper Amending Motions have been submitted to the NFPA by the deadline of May 7, 2010. For more information on the new rules and for up-to-date information on schedules and deadlines for processing NFPA Documents, check the NFPA website (www.nfpa.org) or contact NFPA Standards Administration.

National Fire Protection Association®
1 BATTERYMARCH PARK, QUINCY, MA 02169-7471

ISSN 1079-5332   Copyright © 2010 All Rights Reserved

National Electrical Code, NEC, NFPA, and National Fire Protection Association are registered trademarks of the National Fire Protection Association, Quincy, MA 02169

Figure 1-5. The Report on Comments (ROC) is a compilation of the documented action on comments received during a Code-making cycle.

Appeals to the Standards Council

Appeals to the Standards Council may be made within 20 days after the Annual Meeting, after which the Standards Council adjudicates any appeals, accepts the new *NEC*, and issues the revised *Code*.

NEC COMMITTEES

In addition to the Standards Council, the NFPA utilizes specific committees in the development of the *NEC*. These committees collaborate in a seamless structure to ensure that the newly developed code proposals are best suited throughout all the NFPA codes. The committees also perform technical research to better understand the future needs of public safety within the electrical industry. Some of these committees are maintained as part of the standard code-making process, while other committees may be briefly derived to perform a particular task and are then dissolved shortly afterwards.

Technical Correlating Committee (TCC)

The Technical Correlating Committee reviews all of the *Code*-Making Panel results to ensure that there are no conflicting actions. Their duty is extensive, for they are required to review multiple CMP's ROC reports. Typically, the chairmen of the TCs will advise the TCC to ensure that possible conflicting actions are resolved. See **Figure 1-6.**

Figure 1-6   2011 *NEC* Technical Correlating Committee

James W. Carpenter, *Chair*
International Association of Electrical Inspectors, NC [E]
Rep. International Association of Electrical Inspectors

Mark W. Earley, *Secretary* National Fire Protection Association, MA (nonvoting)	**Jean A. O'Connor,** *Recording Secretary* National Fire Protection Association, MA (nonvoting)

James E. Brunssen, Telcordia, NJ [UT]
  Rep. Alliance for Telecommunications Industry Solutions
	Lawrence S. Ayer, Biz Com Electric, Inc., OH [IM]
	  (Alt. to David L. Hittinger)
	  Rep. Independent Electrical Contractors, Inc.

Merton W. Bunker, Jr., US Department of State, VA [U]
  (VL to Document: 110, Document: 111, Document: 70,
  Document: 70B, Document: 70E, Document: 79, Document:
  790, Document: 791)
	Larry D. Cogburn, Cogburn Bros., Inc., FL [IM]
	  (Alt. to Stanley J. Folz)

James M. Daly, General Cable, NJ [M]
  Rep. National Electrical Manufacturers Association
	James T. Dollard, Jr., IBEW Local Union 98, PA [L]
	  (Alt. to Palmer L. Hickman)
	  Rep. International Brotherhood of Electrical Workers

William R. Drake, Actuant Electrical, CA [M]	**Ernest J. Gallo,** Telcordia Technologies, Inc., NJ [UT]
	  (Alt. to James E. Brunssen)
	  Rep. Alliance for Telecommunications Industry Solutions

Stanley J. Folz, Morse Electric Company, NV [IM]
  Rep. National Electrical Contractors Association
	Daniel J. Kissane, Legrand/Pass & Seymour, NY [M]
	  (Alt. to James M. Daly)
	  Rep. National Electrical Manufacturers Association

Palmer L. Hickman, National Joint Apprentice & Training
  Committee, MD [L]
  Rep. International Brotherhood of Electrical Workers
	Michael E. McNeil, FMC Bio Polymer, ME [U]
	  (Alt. to Danny Liggett)
	  Rep. American Chemistry Council

David L. Hittinger, Independent Electrical Contractors of Greater
  Cincinnati, OH [IM]
  Rep. Independent Electrical Contractors, Inc.
	Mark C. Ode, Underwriters Laboratories Inc., AZ [RT]
	  (Alt. to John R. Kovacik)

John R. Kovacik, Underwriters Laboratories Inc., IL [RT]	**Richard P. Owen,** Oakdale, MN [E]
	  (Alt. to James W. Carpenter)
	  Rep. International Association of Electrical Inspectors

Neil F. LaBrake, Jr., National Grid, NY [UT] Rep. Electric Light & Power Group/EEI	**Nonvoting**

Danny Liggett, DuPont Engineering, Inc., TX [U]
  Rep. American Chemistry Council
	Richard C Biermann, Biermann Electric Company, Inc., MA [IM]
	  (Member Emeritus)

Alternates	**David Mascarenhas,** Canadian Standards Association, Canada [RT]

Thomas L. Adams, Engineering Consultant, IL [UT]
  (Alt. to Neil F. LaBrake, Jr.)
  Rep. Electric Light & Power Group/EEI
	D. Harold Ware, Libra Electric Company, OK [IM]

	Mark W. Earley, NFPA Staff Liaison

Figure 1-6. The Technical Correlating Committee is the watchdog which ensures there are no conflicting actions between CMPs.

Code-Making Panels

Presently, the *NEC* CMPs consist of nineteen panels. As new technologies develop, the Standard Council will recognize the need to derive new CMPs. Later in this textbook you will gain an understanding of which electrical installation and equipment is covered under each of the 19 CMPs. **See Figure 1-7.**

The IBEW and *NECA* are represented on many of the NFPA CMPs. All the *NEC* panels and members are listed in the front of the *NEC*. Likewise, for other NFPA Codes and Standards (i.e. NFPA 72 Fire Alarm), the CMP members are listed in the front of the publication. Take time to review the CMP listing to get a sense who is occupying the CMPs and being the driving force behind the code-making process. For an example, members of 2011 *NEC* CMP-3 are listed in **Figure 1-8.**

Fact

Each *Code*-Making Panel consists of a chairman, principal members, and alternate members. Only principal members may vote in the *NEC* process. The alternate's vote is only counted when the principal member does not return the written ballot.

Figure 1-7 — 2011 *NEC Code*-Making Panels

NEC CODE-MAKING PANEL	ARTICLES, ANNEX AND CHAPTER 9 MATERIAL WITHIN THE SCOPE OF THE CODE-MAKING PANEL
1	90, 100, 110, Chapter 9, Table 10, Annex A, Annex H, Annex I
2	210, 215, 220, Annex D Examples D1 through D6
3	300, 590, 720, 725, 727, 760, Chapter 9, Tables 11(A) and (B), Tables 12(A) and (B)
4	225, 230, 690, 692, 694, 705
5	200, 250, 280, 285
6	310, 400, 402, Chapter 9 Tables 5 through 9, and Annex B
7	320, 322, 324, 326, 328, 330, 332, 334, 336, 338, 340, 382, 394, 396, 398, 399
8	342, 344, 348, 350, 352, 353, 354, 355, 356, 358, 360, 362, 366, 368, 370, 372, 374, 376, 378, 380, 384, 386, 388, 390, 392, Chapter 9, Tables 1 through 4, and Annex C
9	312, 314, 404, 408, 450, 490
10	240
11	409, 430, 440, 460, 470, Annex D, Example D8
12	610, 620, 625, 626, 630, 640, 645, 647, 650, 660, 665, 668, 669, 670, 685, Annex D, Examples D9 and D10
13	445, 455, 480, 695, 700, 701, 702, 708, Annex F, and Annex G
14	500, 501, 502, 503, 504, 505, 506, 510, 511, 513, 514, 515, 516
15	517, 518, 520, 522, 525, 530, 540
16	770, 800, 810, 820, 830, 840
17	422, 424, 426, 427, 680, 682
18	406, 410, 411, 600, 605
19	545, 547, 550, 551, 552, 553, 555, 604, 675, and Annex D, Examples D11 and D12

Figure 1-7. The are 19 CMPs for the NEC.

COMMITTEE MEMBERSHIP

Members of the NFPA Committees are made up of subject-matter experts who are employed or associated with the subject of the committee. As discussed earlier, the committees are made up of volunteer users of the *Code* such as: research/testing labs, enforcing authorities, insurance, consumers, manufacturing, utilities, and special experts.

To become a member the NFPA, the applicant is required to fill out a form which requests a variety of information such as qualifications, relationship to other members, availability to actively participate, funding source for your participation, background of your employer, and other information such as which organization you represent. As you can see, the committee membership is made up of highly qualified personnel with public safety as a high priority.

Classification

The *Code*-Making Panels, also known as the Technical Committees (TCs), are comprised of volunteers. **See Figure 1-8.** A listing of the scope for each *Code*-Mak-

Figure 1-8	2011 *NEC* CMP No. 3 Members

Paul J. Casparro, *Chair*
Scranton Electricians JATC, PA [L]
Rep. International Brotherhood of Electrical Workers

Lawrence S. Ayer, Biz Com Electric, Inc., OH [IM]
    Rep. Independent Electrical Contractors, Inc.

Thomas F. Connaughton, Intertek Testing Services, NJ [RT]

Les Easter, Tyco/Allied Tube and Conduit, IL [M]
    Rep. National Electrical Manufacturers Association

Sanford E. Egesdal, Egesdal Associates PLC, MN [M]
    Rep. Automatic Fire Alarm Association, Inc.

Stanley D. Kahn, Tri-City Electric Company, Inc., CA [IM]
    Rep. National Electrical Contractors Association

Ray R. Keden, ERICO, Inc., CA [M]
    Rep. Building Industry Consulting Services International

Juan C. Menendez, Southern California Edison Company, CA [UT]
    Rep. Electric Light & Power Group/EEI

Richard P. Owen, Oakdale, MN [E]
    Rep. International Association of Electrical Inspectors

Steven J. Owen, Steven J. Owen, Inc., AL [IM]
    Rep. Associated Builders & Contractors

David A. Pace, Olin Corporation, AL [U]
    Rep. American Chemistry Council

Melvin K. Sanders, Things Electrical Co., Inc. (TECo., Inc.), IA [U]
    Rep. Institute of Electrical & Electronics Engineers, Inc.

Mark A. Sepulveda, USA Alarm Systems, Inc., CA [IM]
    Rep. National Burglar & Fire Alarm Association
    (VL to 720, 725, 727, 760)

John F. Sleights, Travelers Insurance Company, CT [I]

Susan L. Stene, Underwriters Laboratories Inc., CA [RT]

Alternates

Richard S. Anderson, RTKL Associates Inc., VA [M]
    (Alt. to Ray R. Keden)
    Rep. Building Industry Consulting Services International

Steven D. Burlison, Progress Energy, FL [UT]
    (Alt. to Juan C. Menendez)
    Rep. Electric Light & Power Group/EEI

Shane M. Clary, Bay Alarm Company, CA [M]
    (Alt. to Sanford E. Egesdal)
    Rep. Automatic Fire Alarm Association, Inc.

Adam D. Corbin, Corbin Electrical Services, Inc., NJ [IM]
    (Alt. to Lawrence S. Ayer)
    Rep. Independent Electrical Contractors, Inc.

Danny Ligget, DuPont Company, TX [U]
    (Alt. to David A. Pace)
    Rep. American Chemistry Council

T. David Mills, Savannah River Nuclear Solutions, LLC, SC [U]
    (Alt. to Melvin K. Sanders)
    Rep. Institute of Electrical & Electronics Engineers, Inc.

Mark C. Ode, Underwriters Laboratories Inc., AZ [RT]
    (Alt. to Susan L. Stene)

Roger S. Passmore, IES Industrial, Inc., SC [IM]
    (Alt. to Steven J. Owen)
    Rep. Associated Builders & Contractors

Marty L. Riesberg, IBEW Local Union 22, MD [L]
    (Alt. to Paul J. Casparro)
    Rep. International Brotherhood of Electrical Workers

George A. Straniero, Tyco/AFC Cable Systems, Inc., NJ [M]
    (Alt. to Les Easter)
    Rep. National Electrical Manufacturers Association

Robert J. Walsh, City of Hayward, CA [E]
    (Alt. to Richard P. Owen)
    Rep. International Association of Electrical Inspectors

Wendell R. Whistler, Intertek Testing Services, OR [RT]
    (Alt. to Thomas F. Connaughton)

Nonvoting

Edward C. Lawry, Oregon, WI [E]
    (Member Emeritus)

Figure 1-8. CMP No. 3 for the NEC has an IBEW chairman from the Scranton Electricians JATC, PA.

ing Panel and the committee list appear in the front of the *NEC*. In the committee list, TC member names are followed by their employer names and identification letter(s). The identification letters appear in brackets, such as [L] for Labor. Most organizations represented have a principal member and an alternate member. Committee membership classification is part of the balancing process of each panel. **See Figure 1-9**.

Representation

Many organizations take part in the *NEC* process and provide representation for the membership classification which applies to their organization. **See Figure 1-11**. For example, the members of the Technical Committees with the classification E, for Enforcing Authority, are representatives of the IAEI, International Association of Electrical Inspectors. **See Figure 1-10**.

Fact

Often times Insurance Underwriters are members of CMPs. The designation for insurance companies on the CMP list is "I."

Figure 1-9	Committee Member Classifications
M	Manufacturers: makers of products affected by the *NEC*
U	Users: users of the *NEC*
I/M	Installers/Maintainers: installers/maintainers of systems covered by the *NEC*
L	Labor: those concerned with safety in the workplace
R/T	Research/Testing Labs: independent organizations developing/reinforcing standards
E	Enforcing Authority: inspectors, enforcers of the *NEC*
I	Insurance: insurance companies, bureaus, or agencies
C	Consumers: purchasers of products/systems not included in "U," users
SE	Special Experts: providers of special expertise, not applicable to other classifications
UT	Utilities: installers/maintainers of systems not covered by *NEC*

Figure 1-9. CMP members come from a wide background of expertise such as manufacturers and research labs.

Figure 1-10	NFPA Electrical Engineering Division Technical Staff

NFPA Electrical Engineering Division Technical Staff

William Burke, Division Manager
Mark W. Earley, Chief Electrical Engineer
Mark Cloutier, Senior Electrical Engineer
Christopher Coache, Senior Electrical Engineer
Jean A. O'Connor, Electrical Project Specialist/Support Supervisor
Lee. F. Richardson, Senior Electrical Engineer
Richard J. Roux, Senior Electrical Specialist
Jeffrey S. Sargent, Senior Electrical Specialist

Support Staff

Carol Henderson
Mary Warren- Pilson
Kimberly Shea

NFPA Staff Editors

Pamela Nolan
Kim Cervantes

Reprinted with permission from NFPA 70-2011, *National Electrical Code*®, Copyright© 2010, National Fire Protection Association, Quincy, MA 02169. This reprinted material is not the complete and official position of the NFPA on the referenced subject, which is represented only by the standard in its entirety.

Figure 1-10. The NFPA staff consist of technical, support, and editor personnel.

Figure 1-11 — Organizations Represented

Air-Conditioning, Heating, & Refrigeration Institute	Instrumentation, Systems, & Automation Society
Alliance for Telecommunications Industry Solutions	Insulated Cable Engineers Association Incorporated
Alliance of Motion Picture and Television Producers	International Association of Electrical Inspectors
The Aluminum Association Incorporated	International Alliance of Theatrical Stage Employees
American Chemistry Council	International Brotherhood of Electrical Workers
American Iron and Steel Institute	International Electrical Testing Association Incorporated
American Institute of Organ Builders	International Sign Association
American Lighting Association	National Association of RV Parks and Campgrounds
American Petroleum Institute	National Burglar & Fire Alarm Association
American Society of Agricultural & Biological Engineers	National Cable & Telecommunications Association
American Society for Healthcare Engineering	National Electrical Contractors Association
American Wind Energy Association	National Electrical Manufacturers Association
Associated Builders and Contractors	National Elevator Industry Incorporated
Association of Higher Education Facilities Officers	Outdoor Amusement Business Association Incorporated
Association of Pool & Spa Professionals	Power Tool Institute Incorporated
Automatic Fire Alarm Association	Satellite Broadcasting & Communications Association
Building Industry Consulting Service International	Society of Automotive Engineers - Hybrid Committee
CSA/Canadian Electrical Code Committee	Society of the Plastics Industry Incorporated
Copper Development Association Incorporated	Solar Energy Industries Association
Electric Light & Power Group/EEI	TC on Airport Facilities
Grain Elevator and Processing Society	TC on Electrical Systems
Illuminating Engineering Society of North America	Telecommunications Industry Association
Independent Electrical Contractors	Transportation Electrification Committee
Information Technology Industry Council	U.S. Institute for Theatre Technology
Institute of Electrical & Electronics Engineers	The Vinyl Institute

Figure 1-11. Many interested organizations are members of Code-Making Panels.

Summary

The *NEC* revision process is open to all members of the public who wish to take part by submitting proposed changes. The many organizations which participate in the *NEC Code*-making process help to build a true consensus code. Since the *NEC* will change every three years; memorizing requirements or sections may be a wasted effort due to the *Code's* three-year revision cycle. The *Codeology* method, however, will not change. Applying the *Codeology* method will lead to the quick and accurate location of necessary information in the *NEC* today and in all future and past versions of the *Code*.

Review Questions

1. Proposals and comments sent to the National Fire Protection Association to change NFPA 70, the *National Electrical Code*, may be submitted by __?__.
    a. organizations represented in the code-making process
    b. large manufacturers of electrical equipment
    c. anyone who is interested
    d. NFPA members

2. Actions taken by the Technical Committee or *Code*-Making Panel require a(n) __?__ vote at the ROP or ROC meeting for passage.
    a. written
    b. Chairman's
    c. two-thirds
    d. simple majority

3. Actions taken by the Technical Committee or *Code*-Making Panel require a(n) __?__ vote on the written ballot of the ROP or ROC for passage.
    a. written
    b. Chairman's
    c. two-thirds
    d. simple majority

4. The *National Electrical Code* is an ANSI document, which means it is developed through a(n) __?__ standards development process.
    a. consensus
    b. private
    c. members-only
    d. governmental

5. The committee membership classification U designates a person representing __?__.
    a. utilities
    b. users
    c. underwriters
    d. the United States government

6. When a comment is submitted containing new material that did not have public review in the ROP, the action taken by the CMP will be to __?__.
    a. accept
    b. reject
    c. accept in principle
    d. hold

7. A panel statement is required on all actions taken in the proposal and comment stages except for an action to __?__.
    a. reject
    b. accept
    c. accept in part
    d. hold

8. The first step in the *NEC* revision process is the __?__.
    a. proposal stage
    b. comment stage
    c. NFPA annual meeting
    d. appeals to the Standards Council

Table of Contents and Basic Arrangement

Chapter 2 of this book introduces the table of contents and chapter arrangements as the beginning step to learning the *Codeology* method. Throughout this text, these basic subdivisions will be repeated time and time again until they become a natural part of the thought process when working with the *NEC*. The *NEC*, along with any other type of installation code, typically begins with a table of contents placed in the front of the document, designed to quickly take users to the information they desire.

Objectives

» Demonstrate familiarity with the *NEC* table of contents and the articles contained within each chapter.

» Identify the purpose of Article 90 as the introduction to the *NEC* which outlines basic information and requirements.

» Explain how **Section 90.3** describes the arrangement of the *NEC*.

» Identify Chapters 1 through 4 as the chapters which apply generally to all electrical installations.

» Identify Chapters 5, 6, and 7 as special entities which supplement or modify Chapters 1 through 4.

» Identify Chapter 8 as the chapter which stands alone and that Chapters 1 through 7 are to be applied to Chapter 8 installations only when referenced in Chapter 8.

» Describe when to apply the tables found in Chapter 9.

» Describe the purpose of the Informative Annexes.

Chapter **2**

Table of Contents

NEC TABLE OF CONTENTS

Structure of the Table of Contents

The table of contents (TOC) in the *National Electrical Code* is the starting point for all of our inquiries. Once a need or question arises for information available in the *NEC*, the first step is to use the table of contents to reach the right chapter, article, and part. Thus, it is essential that all users of the *NEC* be familiar with the TOC and its contents.

The table of contents is comprised of ten separate sections. The first is the introduction to the *NEC*, **Article 90**. The remaining subdivisions are divided into nine chapters, each covering broad areas. These chapters are then subdivided into articles and parts which will be explored later in this text. The introduction (**Article 90**) and each of the nine chapters will also be covered in depth later in this text.

> **Fact**
>
> The table of contents is always the starting point when using the *NEC*.

Article 90 Introduction
Chapter 1 General
Chapter 2 Wiring and Protection
Chapter 3 Wiring Methods and Materials
Chapter 4 Equipment for General Use
Chapter 5 Special Occupancies
Chapter 6 Special Equipment
Chapter 7 Special Conditions
Chapter 8 Communications Systems
Chapter 9 Tables and Informative Annexes

2011 *NEC* Table of Contents

The NFPA 70 *NEC* is revised every three years to facilitate new technologies and to improve public safety. The TOC continues to expand with new sections and parts which the user must review. **See Figure 2-1.**

Figure 2-1   2011 *NEC* Table of Contents

Figure 2-1. The 2011 NEC TOC lists an introduction (Article 90) and nine chapters.

ARRANGEMENT OF THE *NEC*

Article 90 Introduction

The first major subdivision of the *NEC* is **Article 90 Introduction**. It provides the basic information and requirements necessary to properly apply the rest of the document. **Article 90** provides the ground rules upon which the rest of the *NEC* is based. In laying the ground rules, the following sections of **Article 90** establish the format of the *Code*.

Section 90.1 Purpose
Section 90.2 Scope
Section 90.3 Code Arrangement

Section 90.3 Code Arrangement - One of the most basic, yet extremely important facts about the structure of the *NEC* is that its table of contents is specifically designed to facilitate the proper application of each chapter. **Article 90 Introduction** details this arrangement in **Section 90.3** by illustrating the division of the *NEC* into the introduction and nine chapters. **See Figure 2-2**. **Section 90.1** and **Section 90.2** will be discussed in detail in later chapters.

As required in **Section 90.3**, Chapters 1 through 4 apply generally in all electrical installations. They contain the basic electrical installation requirements for all electrical installations, from a single-family dwelling unit to a petroleum refinery or a hospital. A comprehensive understanding of these four chapters is imperative because they are the backbone for all electrical installations.

Chapter 6 of this text explores the remaining **Sections 90.4 – 90.9** in detail. These sections set forth enforcement, rules, Informative Annexes, formal interpretations, safety examinations, and wiring planning.

Fact

The nine chapters of the *NEC* logically separate installation requirements to aid the *Code* user to quickly locate needed information.

Figure 2-2	Chapter Arrangement of the *NEC*
Article 90	Introduction
Chapters 1 through 4 apply GENERALLY to ALL electrical installations.	
Chapter 1	General
Chapter 2	Wiring and Protection
Chapter 3	Wiring Methods and Materials
Chapter 4	Equipment for General Use
Chapters 5, 6, and 7 SUPPLEMENT or MODIFY Chapters 1 through 4.	
Chapter 5	Special Occupancies
Chapter 6	Special Equipment
Chapter 7	Special Conditions
Chapters 1 through 7 DO NOT apply to Chapter 8 unless there is a specific reference in Chapter 8 referring to another chapter.	
Chapter 8	Communications Systems
The tables in Chapter 9 apply as referenced elsewhere in the *NEC*®.	
Chapter 9	Tables
Informative Annexes A through I are for informational purposes only and are not mandatory.	
Chapter 9	Informative Annexes A through I

Figure 2-2. Section 90.3 describes the chapter arrangements of the NEC.

Chapter 1 General

Although the title of Chapter 1 is simply "General", its scope is general information and rules for electrical installations, and pertains to all chapters of the *NEC*.

Chapter 1 consists of 2 Articles:
100 Definitions
110 Requirements for Electrical Installations

Chapter 2 Wiring and Protection

The title of Chapter 2 is Wiring and Protection. However, its scope is information and rules on wiring and protection of electrical installations. Articles 200 through 230 address "wiring" and include information related to grounded conductors, calculations for conductor size, branch circuits, feeders, and services. Articles 240 through 285 address "protection" and include information related to the use of overcurrent protection, grounding, bonding, surge arresters and SPDs. **See Figure 2-3.**

Chapter 2 consists of 10 Articles:
200 Use and Identification of Grounded Conductors
210 Branch Circuits
215 Feeders
220 Branch-Circuit, Feeder, and Service Calculations
225 Outside Branch Circuits and Feeders
230 Services
240 Overcurrent Protection
250 Grounding and Bonding
280 Surge Arresters, Over 1 kV
285 Surge-Protective Devices (SPDs), 1 kV or Less

Chapter 3 Wiring Methods and Materials

The scope of Chapter 3 is information and rules on wiring methods and materials for use in electrical installations. Its articles include information on all permitted methods and materials to supply an electrical installation. This chapter details requirements for all wiring methods and materials from the service point to termination at the last outlet in the electrical distribution system. It includes, for example, general information for all wiring methods, types of cable assemblies, types of raceways, cabinets, cutout boxes, meter socket enclosures, types of boxes, conduit bodies, fittings, and more.

Chapter 3 consists of 45 Articles:
300 Wiring Methods
310 Conductors for General Wiring
312 Cabinets, Cutout Boxes, and Meter Socket Enclosures
314 Outlet, Device, Pull, and Junction Boxes; Conduit Bodies; Fittings; and Handhole Enclosures
320 Armored Cable: Type AC
322 Flat Cable Assemblies: Type FC
324 Flat Conductor Cable: Type FCC
326 Integrated Gas Spacer Cable: Type IGS
328 Medium Voltage Cable: Type MV
330 Metal-Clad Cable: Type MC
332 Mineral-Insulated, Metal-Sheathed Cable: Type MI
334 Nonmetallic-Sheathed Cable: Types NM, NMC, and NMS
336 Power and Control Tray Cable: Type TC

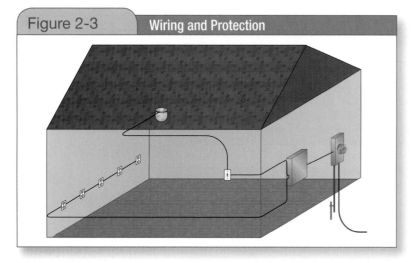

Figure 2-3 | **Wiring and Protection**

Figure 2-3. Chapter 2 Wiring and Protection covers the feeders, branch circuits, services, and grounding for most buildings.

Chapter 4 Equipment for General Use

The scope of Chapter 4 is information and rules on equipment for general use in electrical installations. Articles in this chapter address equipment for general use and include information on all equipment in an electrical installation. Note that Chapter 3 addresses wiring methods and materials, whereas Chapter 4 addresses all electrical equipment necessary for utilization, control, generation, and transformation of electrical energy in an electrical installation. For example, Chapter 4 includes requirements for equipment cords/cables, switches, receptacles, panelboards, generators, transformers, appliances, motors, and other utilization equipment. **See Figure 2-4.**

<u>Chapter 4 consists of 21 Articles:</u>

400 Flexible Cords and Cables

402 Fixture Wires

404 Switches

406 Receptacles, Cord Connectors, and Attachment Plugs (Caps)

408 Switchboards and Panelboards

409 Industrial Control Panels

410 Luminaires, Lampholders, and Lamps

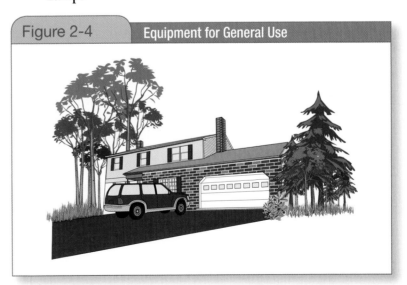

| Figure 2-4 | Equipment for General Use |

Figure 2-4. Chapter 4 covers the requirements for switches, receptacles, luminaries, appliances, motors, generators, and many other typical components within a building.

411 Lighting Systems Operating at 30 Volts or Less

422 Appliances

424 Fixed Electric Space-Heating Equipment

426 Fixed Outdoor Electric Deicing and Snow-Melting Equipment

427 Fixed Electric Heating Equipment for Pipelines and Vessels

430 Motors, Motor Circuits, and Controllers

440 Air-Conditioning and Refrigerating Equipment

445 Generators

450 Transformers and Transformer Vaults (Including Secondary Ties)

455 Phase Converters

460 Capacitors

470 Resistors and Reactors

480 Storage Batteries

490 Equipment Over 600 Volts, Nominal

Note

The requirements of Chapters 1 through 4 apply to all electrical installations, including one-family dwellings to multifamily dwellings. These requirements also apply to the electrical installation in commercial occupancies such as food markets, drug stores, bakeries, shopping malls, or office spaces.

SPECIAL CHAPTERS

As required in **Section 90.3**, Chapters 5, 6, and 7 supplement or modify the first four chapters. Unlike the first four, which are general for all electrical installations, Chapters 5, 6, and 7 cover special requirements. They apply to special occupancies, special equipment, or special conditions.

Chapter 5 Special Occupancies

The scope of this chapter is modifications and/or supplemental information and rules for electrical installations in special occupancies. As stated in **Section 90.3** (**see Figure 2-2**), this chapter includes "occupancy-specific" information supplementing or modifying the first four chapters in special occupancies such as

Fact

Chapter 5 modifies and supplements the general rules of Chapters 1 through 4 for special occupancies.

hazardous locations, health care facilities, places of assembly, carnivals, agricultural buildings, mobile homes, RVs, and marinas. **See Figure 2-5.**

Chapter 5 consists of 28 Articles:

500 Hazardous (Classified) Locations, Classes I, II, and III, Divisions 1 and 2

501 Class I Locations

502 Class II Locations

503 Class III Locations

504 Intrinsically Safe Systems

505 Zone 0, 1, and 2 Locations

506 Zone 20, 21, and 22 Locations for Combustible Dusts, or Ignitible Fibers/ Flyings

510 Hazardous (Classified) Locations - Specific

511 Commercial Garages, Repair and Storage

513 Aircraft Hangars

514 Motor Fuel Dispensing Facilities

515 Bulk Storage Plants

516 Spray Application, Dipping and Coating Processes

517 Health Care Facilities

518 Assembly Occupancies

520 Theatres, Audience Areas of Motion Picture and Television Studios, Performance Areas, and Similar Locations

522 Control Systems for Permanent Amusement Attractions

525 Carnivals, Circuses, Fairs, and Similar Events

530 Motion Picture and Television Studios and Similar Locations

540 Motion Picture Projection Rooms

545 Manufactured Buildings

547 Agricultural Buildings

550 Mobile Homes, Manufactured Homes, and Mobile Home Parks

551 Recreational Vehicles and Recreational Vehicle Parks

552 Park Trailers

553 Floating Buildings

555 Marinas and Boatyards

590 Temporary Installations

Note

The requirements in Chapters 1 through 4 apply to electrical installations in special occupancies such as hospitals, marinas, refineries, industries, farms, and the like, with the requirements of Chapter 5 supplementing or modifying those rules.

Chapter 6 Special Equipment

The scope of Chapter 6 is modifications and/or supplemental information for electrical installations containing special equipment. This chapter includes "equipment-specific" information supplementing or modifying the first four chapters for special equipment, such as electric signs, welders, x-ray equipment, swimming pools, solar photovoltaic systems, fuel cells, and fire pumps. **See Figure 2-6.**

Chapter 6 consists of 25 Articles:
  600 Electric Signs and Outline Lighting
  604 Manufactured Wiring Systems
  605 Office Furnishings (Consisting of Lighting Accessories and Wired Partitions)
  610 Cranes and Hoists
  620 Elevators, Dumbwaiters, Escalators, Moving Walks, Platform Lifts, and Stairway Chair Lifts
  625 Electric Vehicle Charging System
  626 Electrified Truck Parking Space
  630 Electric Welders
  640 Audio Signal Processing, Amplification, and Reproduction Equipment
  645 Information Technology Equipment
  647 Sensitive Electronic Equipment
  650 Pipe Organs
  660 X-Ray Equipment
  665 Induction and Dielectric Heating Equipment
  668 Electrolytic Cells
  669 Electroplating
  670 Industrial Machinery

  675 Electrically Driven or Controlled Irrigation Machines
  680 Swimming Pools, Fountains, and Similar Installations
  682 Natural and Artificially Made Bodies of Water

Figure 2-5   Special Occupancies

Figure 2-5. Chapter 5 covers the requirements special occupancies such as hospitals.

Fact

Chapter 6 modifies and supplements the general rules of Chapters 1 through 4 for special equipment.

Figure 2-6   Special Equipment

Figure 2-6. Chapter 6 covers the requirements for special equipment such as swimming pools and spas.

685 Integrated Electrical Systems
690 Solar Photovoltaic Systems
692 Fuel Cell Systems
694 Small Wind Electric Systems
695 Fire Pumps

Note

The requirements of Chapter 6 supplement and/or modify Chapters 1 through 4 to accommodate specific requirements for an electrical installation with special equipment. Examples of special equipment include electric signs, welders, X-ray equipment, swimming pools, solar photovoltaic systems, fuel cells, and fire pumps.

Chapter 7 Special Conditions

The scope of Chapter 7 is modifications and/or supplemental information for electrical installations containing special conditions. This chapter includes "condition-specific" information supplementing or modifying the first four chapters for special conditions, such as emergency systems, legally-required standby systems, Class 1, 2, and 3 systems, and fire alarm systems. **See Figure 2-7.**

Fact

Chapter 7 modifies and supplements the general rules of Chapter 1 through 4 for special conditions.

Chapter 7 consists of 10 Articles:
700 Emergency Systems
701 Legally Required Standby Systems
702 Optional Standby Systems
705 Interconnected Electric Power Production Sources
708 Critical Operations Power Systems (COPS)
720 Circuits and Equipment Operating at Less Than 50 Volts
725 Class 1, Class 2, and Class 3 Remote-Control, Signaling, and Power-Limited Circuits
727 Instrumentation Tray Cable: Type ITC
760 Fire Alarm Systems
770 Optical Fiber Cables and Raceways

Note

The requirements of Chapter 7 supplement and modify Chapters 1 through 4 to accommodate specific requirements for an electrical installation with special conditions. Examples of special conditions include emergency systems, legally required standby systems, Class 1 systems, Class 2 systems, Class 3 systems, and fire alarms.

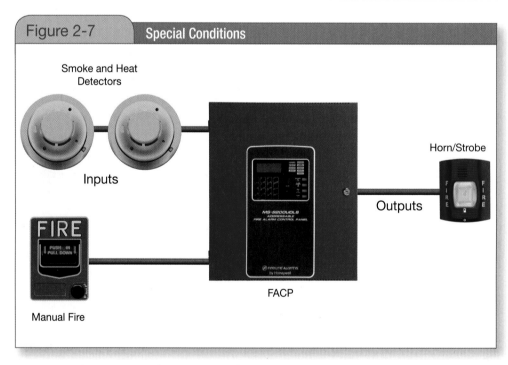

Figure 2-7 **Special Conditions**

Smoke and Heat Detectors

Inputs

Horn/Strobe

Outputs

FIRE

FACP

Manual Fire

Figure 2-7. Chapter 7 covers the requirements for special conditions such as emergency equipment such as Fire Alarm Systems.

Chapter 8 Communications Systems

Chapter 8 covers communications systems and is not subject to the requirements of Chapters 1 through 7 (as per Section 90.3) except where the requirements are specifically referenced in Chapter 8. **See Figure 2-2.** This means that all of the Articles listed in Chapter 8 stand alone and are not subject to the rules in the rest of the *NEC* unless a Chapter 8 Article specifically references a requirement elsewhere in the *Code*. Chapter 8 includes specific information for communications systems, such as communications circuits, radio equipment, television equipment, CATV, and broadband systems that are both network powered and all premises-powered broadband communication systems. **See Figure 2-8.**

Chapter 8 consists of 5 Articles:
- 800 Communications Circuits
- 810 Radio and Television Equipment
- 820 Community Antenna Television and Radio Distribution Systems
- 830 Network-Powered Broadband Communications Systems
- 840 Premises-Powered Broadband Communications Systems

Chapter 8 Communications Systems covers installation of coaxial and other radio conductors.

Figure 2-8   Communication Systems

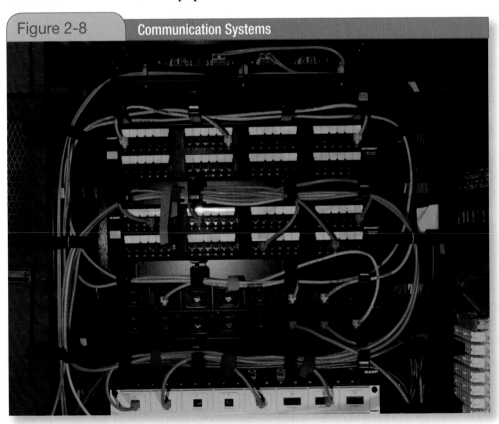

Figure 2-8. Chapter 8 Communications Systems covers systems such as telephone and network racks located in dedicated rooms in a facility.

Chapter 9 Tables

Chapter 9 contains tables which are referenced throughout the *NEC*. **See Figure 2-2.** As per **Section 90.3**, Chapter 9 tables are applicable as referenced. Informative Annexes which follow are not part of the requirements of the *NEC*, but are included for informational purposes. **See Figure 2-9.**

<u>Chapter 9 consists of Tables and Informative Annexes:</u>

Table 1 Percent of Cross Section of Conduit and Tubing for Conductors

Table 2 Radius of Conduit and Tubing Bends

Table 4 Dimensions and Percent Area of Conduit and Tubing (Areas of Conduit or Tubing for the Combinations of Wires Permitted in Table 1, Chapter 9)

Table 5 Dimensions of Insulated Conductors and Fixture Wires

Table 5A Compact Copper and Aluminum Building Wire Nominal Dimensions and Areas

Table 8 Conductor Properties

Table 9 Alternating-Current Resistance and Reactance for 600-Volt Cables, 3-Phase, 60 Hz, 75°C (167°F) -Three Single Conductors in Conduit

Table 10 Conductor Stranding

Table 11A Class 2 & Class 3 Alternating-Current Power Source Limitations

Table 11B Class 2 & Class 3 Direct-Current Power Source Limitations

Table 12A PLFA Alternating-Current Power Source Limitations

Table 12B PLFA Direct-Current Power Source Limitations

Informative Annex A Product Safety Standards

Informative Annex B Application Information for Ampacity Calculation

Informative Annex C Conduit and Tubing Fill Tables for Conductors and Fixture Wires of the Same Size

Informative Annex D Examples

Informative Annex E Types of Construction

Informative Annex F Availability and Reliability for Critical Operations Power Systems; and Development and Implementation of Functional Performance Tests (FPTs) for Critical Operations Power Systems

Informative Annex G Supervisory Control and Data Acquisition (SCADA)

Informative Annex H Administration and Enforcement

Informative Annex I Recommended Tightening Torque Tables from UL Standard 486A-B

Conductor rating in earth covered duct banks is part of Informative Annex B.

Figure 2-9 — Table 8 Conductor Properties (partial)

	Conductors								Direct-Current Resistance At 75°C (167°F)					
	Area			**Stranding**		**Overall**		**Area**		**Copper**				**Aluminum**
				Diameter		**Diameter**				**Uncoated**		**Coated**		
Size (AWG or kcmil)	mm²	Circular mils	Quantity	mm	in.	mm	in.	mm²	in.²	ohm/ km	ohm/ kFT	ohm/ km	ohm/ kFT	ohm/ km	ohm/ kFT
18	0.823	1620	1	—	—	1.01	0.040	0.823	0.001	25.5	7.77	26.5	8.08	42.0	12.8
18	0.823	1620	7	0.39	0.015	1.16	0.046	1.06	0.002	26.1	7.95	27.7	8.45	42.8	13.1
16	1.31	2580	1	—	—	1.29	0.051	1.31	0.002	16.0	4.89	16.7	5.08	26.4	8.05
16	1.31	2580	7	0.49	0.019	1.46	0.058	1.68	0.003	16.4	4.99	17.3	5.29	26.9	8.21
14	2.08	4110	1	—	—	1.63	0.064	2.08	0.003	10.1	3.07	10.4	3.19	16.6	5.06
14	2.08	4110	7	0.62	0.024	1.85	0.073	2.68	0.004	10.3	3.14	10.7	3.26	16.9	5.17
12	3.31	6530	1	—	—	2.05	0.081	3.31	0.005	6.34	1.93	6.57	2.01	10.45	3.18
12	3.31	6530	7	0.78	0.030	2.32	0.092	4.25	0.006	6.50	1.98	6.73	2.05	10.69	3.25
10	5.261	10380	1	—	—	2.588	0.102	5.26	0.008	3.984	1.21	4.148	1.26	6.561	2.00
10	5.261	10380	7	0.98	0.038	2.95	0.116	6.76	0.011	4.070	1.24	4.226	1.29	6.679	2.04
8	8.367	16510	1	—	—	3.264	0.128	8.37	0.013	2.506	0.764	2.579	0.786	4.125	1.26
8	8.367	16510	7	1.23	0.049	3.71	0.146	10.76	0.017	2.551	0.778	2.653	0.809	4.204	1.28
6	13.30	26240	7	1.56	0.061	4.67	0.184	17.09	0.027	1.608	0.491	1.671	0.510	2.652	0.808
4	21.15	41740	7	1.96	0.077	5.89	0.232	27.19	0.042	1.010	0.308	1.053	0.321	1.666	0.508
3	26.67	52620	7	2.20	0.087	6.60	0.260	34.28	0.053	0.802	0.245	0.833	0.254	1.320	0.403

Figure 2-9. Table 8 Conductor Properties is utilized often when calculating the conduit size for a quantity of different size conductors.

Summary

The starting point for all inquiries in the *NEC* is the table of contents (TOC). The *NEC* is subdivided into ten major subdivisions:

- Article 90 Introduction
- Chapter 1 General
- Chapter 2 Wiring and Protection
- Chapter 3 Wiring Methods and Materials
- Chapter 4 Equipment for General Use
- Chapter 5 Special Occupancies
- Chapter 6 Special Equipment
- Chapter 7 Special Conditions
- Chapter 8 Communications Systems
- Chapter 9 Tables and Informative Annexes

One of the most basic and crucial steps toward being proficient in the use of the *National Electrical Code* is an in-depth understanding of how the *Code* is arranged. In **Article 90**, the introduction to the *NEC*, **Section 90.3** explains the *Code* arrangement in detail as follows:

- The *NEC* is divided into the introduction and nine chapters.
- Chapters 1, 2, 3, and 4 apply generally to all electrical installations.
- Chapters 5, 6, and 7 are special chapters which apply to special occupancies, special equipment, and special conditions. These special chapters supplement and/or modify the general rules in Chapters 1 through 4.
- Chapter 8 stands alone and is not subject to the requirements of Chapters 1 through 7 unless specifically referenced in a Chapter 8 Article.
- Chapter 9 contains tables which are applicable only when referenced.
- Informative Annexes are informational only and are not mandatory.

An in-depth understanding of this arrangement is necessary to properly apply the requirements of the *NEC*. Understanding the arrangement of the *Code* and using the table of contents as a starting point for all inquiries into the *NEC* are the cornerstones of the *Codeology* method.

Review Questions

1. **The table of contents is broken down into __?__ major subdivisions.**
   a. 10
   b. 8
   c. 2
   d. 9

2. **The Introduction to the *NEC* is located in __?__ .**
   a. preface
   b. TOC
   c. Chapter 1
   d. Article 90

3. Articles within the *NEC* in the 500 series are dedicated in scope to __?__.
   a. utilization equipment
   b. special occupancies
   c. special conditions
   d. wiring methods

4. Which chapter will contain an Article to address the special equipment requirements for swimming pools?
   a. 8
   b. 7
   c. 6
   d. 5

5. The answer to a question about the size of conductors supplying a motor in a general application will be found in Chapter __?__.
   a. 1
   b. 2
   c. 3
   d. 4

6. The tables located in Chapter 9 of the *NEC* apply __?__.
   a. at all times
   b. only in Chapters 5, 6, and 7
   c. wherever they are useful
   d. as referenced in the *NEC*

7. Which chapter of the *NEC* stands alone and is not subject to any other of its chapters unless a specific reference is made?
   a. Chapter 1
   b. Chapter 5
   c. Chapter 8
   d. Chapter 9

8. Chapters 1 through 4 of the *NEC* apply __?__ to all electrical installations.
   a. without modification
   b. generally
   c. sparingly
   d. in some cases

9. Annexes in Chapter 9 of the *NEC* are __?__.
   a. informational only
   b. mandatory requirements
   c. applicable as referenced
   d. used only for special equipment

10. Chapters 5, 6, and 7 of the *NEC* __?__ Chapters 1 through 4.
    a. are not be associated in any way with
    b. do apply generally with
    c. do not have an effect on
    d. supplement or modify

11. The arrangement of the *NEC* is outlined in __?__ of the *Code*.
    a. 90.1
    b. 110.3(B)
    c. 210.8
    d. 90.3

NEC Organization

This chapter covers the *NEC* organization, outline form, and rules which are essential for the understanding and application of the *Codeology* method. Throughout this text, these basic principles are repeated time and time again until they become part of the natural thought process when working with this *Code*. The *NEC* structure is governed and outlined by the *National Electrical Code Style Manual*.

Objectives

» Outline the structure of the *NEC* from chapters to articles, parts, sections, three levels of subdivisions, exceptions, and list items.

» Identify the mandatory and permissive text within the *NEC*.

» Recognize that informational notes are informational only.

» Recognize the location and importance of cross-reference tables.

» Recognize the usefulness of outlines, diagrams, and drawings within the *NEC*.

» Identify the application of informative annexes, units of measurement, and extract material.

Chapter 3

Table of Contents

NEC ORGANIZATION

The *National Electrical Code* is a well-organized installation document. The rules which govern its structure, known as the *National Electrical Code Style Manual*, are available online at www.nfpa.org from the *National Fire Protection Agency*. The *NEC Style Manual* is intended to be used as a practical working tool to assist in making the *Code* as clear, usable, and unambiguous as possible. Keep in mind the *NEC Style Manual* is structured differently than the *NFPA Manual of Style* used in other NFPA Codes. **See Figure 3-1.** In examining the *NEC* structure, the starting point is the Table of Contents.

OUTLINE FORM OF THE *NEC*

Chapters

The Table of Contents is sectioned into 10 separate areas. The first, **Article 90**, introduces the *NEC*. It is then subdivided into chapters which are major subdivisions covering broad areas divided into articles. Nine chapters follow the Introduction; the first eight outlining a number of articles under the title and scope of the chapter. Chapter 9 is comprised of tables and informative annexes.

Articles

Articles are subdivisions of chapters which cover specific subjects such as branch circuits, grounding, transformers, rigid metal conduit, motors, and the like. Each article is given an individual title and divided into sections and sometimes parts. For example, Chapter 1 is titled General, and the articles within this chapter must be of a general nature for all electrical installations. The articles contained in Chapter 1, the 100 series, are general in nature and are titled, **Article 100 Definitions** and **Article 110 Requirements for Electrical Installations**. **See Figure 3-2.**

When sufficiently large, an article is sometimes subdivided into parts and sections, which correspond to logical groupings of information.

Parts

Parts are given individual titles designated by Roman numerals. When necessary, due to large size or for usability purposes, an article is subdivided into separate parts, each dedicated to a logical separation of requirements within the given article. For example, **Chapter 1 General** contains **Article 110 Requirements for Electrical Installations**. This article is then logically separated into five parts which are individually titled to describe the requirements contained within a part. Parts are always numbered with Roman numerals. **Article 110**, for example, has five parts:

- I.   **General**
- II. **600 Volts, Nominal, or Less**
- III. **Over 600 Volts, Nominal**
- IV. **Tunnel Installations over 600 Volts, Nominal**
- V.   **Manholes and Other Electric Enclosures Intended for Personnel Entry, All Voltages**

For additional information, visit qr.njatcdb.org Item #1062

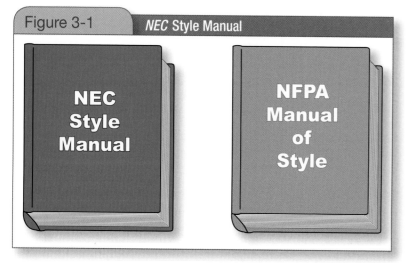

Figure 3-1   *NEC* Style Manual

NEC Style Manual

NFPA Manual of Style

Figure 3-1. Most of the NFPA Codes utilize rules from the NFPA Manual of Style. The NEC (NFPA 70) uses a set of rules specific to the NEC called the NEC Style Manual.

Sections

Each part is then subdivided into individual sections, each dedicated to a separate rule under the title of the specific part. These logical separations of requirements into sections are individually titled to identify the rule(s) and are designed to enable the *Code* user to quickly find the necessary information.

Sections are sometimes divided into up to three levels of subdivisions to clarify a requirement. Sections and the first two levels of subdivisions are always given a title. Sections and subdivisions may contain lists, exceptions, and informational notes. Sections are always identified by a section number and a title in bold print. The following is an explanation of how sections are organized and presented:

110.14 Electrical Connections – *Sections* are always identified by a section number and a title in bold print.

(C) Temperature Limitations – *First-level subdivisions* is always identified by an uppercase letter in parentheses and a title in bold print.

(1) Equipment Provisions – *Second-level subdivisions* are identified by a number in parentheses and a title in bold print.

(a) Termination provisions… – *Third-level subdivisions* are identified by a lowercase letter in parentheses, may be titled, and a title in bold print:

(a) Termination provisions of equipment for circuits rated 100 amperes or less, or marked for 14 AWG through 1 AWG conductors, shall be used only for one of the following:

(1) Conductors rated 60°C (140°F) – *List items* are identified by a lowercase letter or number in parentheses followed by the text, without title, not in bold print.

Figure 3-2 | **NEC Structure Breakdown**

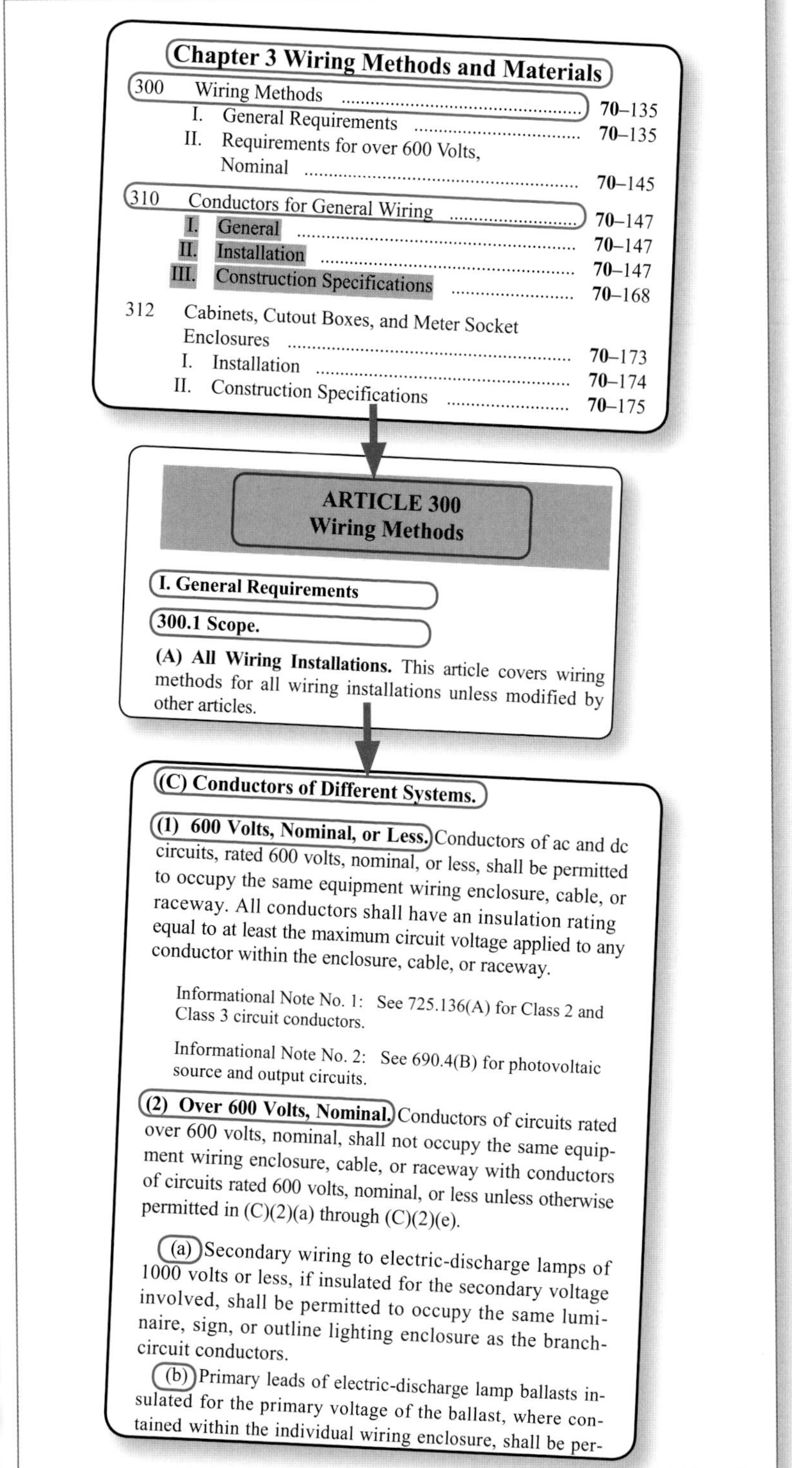

Chapter 3 Wiring Methods and Materials

300	Wiring Methods	**70**–135
	I. General Requirements	**70**–135
	II. Requirements for over 600 Volts, Nominal	**70**–145
310	Conductors for General Wiring	**70**–147
	I. General	**70**–147
	II. Installation	**70**–147
	III. Construction Specifications	**70**–168
312	Cabinets, Cutout Boxes, and Meter Socket Enclosures	
	I. Installation	**70**–174
	II. Construction Specifications	**70**–175

ARTICLE 300 Wiring Methods

I. General Requirements

300.1 Scope.

(A) All Wiring Installations. This article covers wiring methods for all wiring installations unless modified by other articles.

(C) Conductors of Different Systems.

(1) 600 Volts, Nominal, or Less. Conductors of ac and dc circuits, rated 600 volts, nominal, or less, shall be permitted to occupy the same equipment wiring enclosure, cable, or raceway. All conductors shall have an insulation rating equal to at least the maximum circuit voltage applied to any conductor within the enclosure, cable, or raceway.

Informational Note No. 1: See 725.136(A) for Class 2 and Class 3 circuit conductors.

Informational Note No. 2: See 690.4(B) for photovoltaic source and output circuits.

(2) Over 600 Volts, Nominal. Conductors of circuits rated over 600 volts, nominal, shall not occupy the same equipment wiring enclosure, cable, or raceway with conductors of circuits rated 600 volts, nominal, or less unless otherwise permitted in (C)(2)(a) through (C)(2)(e).

(a) Secondary wiring to electric-discharge lamps of 1000 volts or less, if insulated for the secondary voltage involved, shall be permitted to occupy the same luminaire, sign, or outline lighting enclosure as the branch-circuit conductors.

(b) Primary leads of electric-discharge lamp ballasts insulated for the primary voltage of the ballast, where contained within the individual wiring enclosure, shall be per-

Figure 3-2. The NEC structure starts at the Table of Contents then is broken down into Chapters, Articles, Parts (Roman numerals), Sections (XXX.X), and Subdivisions (A-B-C, 1-2-3, a-b-c, etc.).

Exceptions are options identified by italicized text, are not in bold print, and are numbered where more than one exists (i.e. Exception No. 1). Where exceptions are made to items within a numbered list, the exception shall clearly indicate the items within the list to which it applies. An exception immediately follows the section, subdivision, or list item to which it applies.

Informational Notes (previously called Fine Print Notes - FPNs) are followed by informative text. An Informational Note immediately follows the section, subdivision, or list item to which it applies. These notes are informational only and not an enforceable part of the *NEC*:

> Informational Note: With respect to 110.14(C)(1) and (2), equipment markings or listing information may additionally restrict the sizing and temperature ratings of connected conductors.

Tables

Tables, which are located in Chapter 9, are applicable only when referenced in other chapters of the *NEC*. For example, Table 1 is used for "Conduit Fill" and is applicable only where referenced in another section of the *NEC*. Table 1 in Chapter 9 will be permitted for use to calculate raceway fill for raceway articles, such as **Article 358 Electrical Metallic Tubing: Type EMT. Article 358** is subdivided into three parts. Part II is titled "Installation" and includes **Section 358.22**, which references Table 1 in Chapter 9. This specific reference allows the use of the table:

> **358.22 Number of Conductors.** The number of conductors shall not exceed that permitted by the percentage fill specified in Table 1, Chapter 9.
>
> Cables shall be permitted to be installed where such use is not prohibited by the respective cable articles. The number of cables shall not exceed the allowable percentage fill specified in Table 1, Chapter 9.

Informative Annexes

As stated in **Section 90.3**, Informative Annexes are for informational purposes only. Annex material is provided to aid the user in understanding and applying the requirements in the *NEC*. For example, **Informative Annex A - Product Safety Standards** aids the *Code* user by providing a list of product safety standards used for product listing where that listing is required by the *NEC*. **Informative Annex A** is not a part of the requirements of the *Code*, but is included for informational purposes only.

For example, in **Article 348 Flexible Metal Conduit: Type FMC, I. General** contains **Section 348.6 Listing Requirements**, which requires all FMC and associated fittings to be listed. **Informative Annex A** provides the "Product Standard Number" for Flexible Metal Conduit, which is "UL 1."

RULES, EXCEPTIONS, AND INFORMATIONAL NOTES

Application of Rules

The structure of the *NEC* is in a progressive, ladder-type format. **See Figure 3-3.** For example, a rule which exists in a third-level subdivision applies only under the second-level subdivision. In addition, the second-level subdivision is limited to the rule in the first-level subdivision, which applies only under the section in which it exists, which applies only in the part it is arranged in, which applies only in the article in which it exists, which is limited to the chapter in which it is located.

The key to properly applying the rules of the *NEC* is always to apply the rule within the part of the article in which it exists. Without an understanding of the outline form of the *NEC*, a new or inexperienced user might attempt to broadly apply a rule to areas in which it may not apply. Using the *Codeology* method, the user will always know in which part of

Figure 3-3	*NEC* Application of Rules
Section	Applies only within the scope of the part of the article in which it is located.
First-Level Subdivision	Applies only within the scope of the section in which it exists.
Second-Level Subdivision	Applies only within the scope of the first-level subdivision in which it exists.
Third-Level Subdivision	Applies only within the scope of the second-level subdivision in which it exists.
List Items	Applies only in the section or subdivision in which they exist.
Exceptions	Applies only to the section, subdivision, or list item under which they exist.
Informational Notes	Used for informational purposes only and are designed to aid the user in the application of the rule(s) under which they exist.

Figure 3-3. Applications of Rules in the NEC is in a progressive ladder-type format.

what article and chapter the section exists. This basic information is crucial to the proper application of all *NEC* rules.

Exceptions

Exceptions are used only where absolutely necessary. Exceptions are always italicized in the *Code* for quick and easy identification. The *NEC Code*-Making Panels strive to make the *Code* as user-friendly as possible. Over time, most exceptions have been eliminated and the rules changed to positive text, making them easier to read and apply. Exceptions are used in the following cases:

1. The rule in which the exception is applied is modified or supplemented elsewhere in the *NEC*. This modification or supplement will be qualified to very specific locations, equipment, conditions, or wiring methods or uses. Examples of this type of exception include the following:
   - **210.8(A)(3) Exception**
   - **314.28 Exception**
   - **450.3(B) Exception**
2. Existing conditions may require alternative methods or modification of the rule. Examples of this type of

exception include the following:
   - **110.26(E) Exception**
   - **300.20(A) Exception No.1**
3. Exceptions are written to allow specific variations from the general rule to explain and clarify the intent and scope of the rule. Examples of this type of exception follow:
   - **110.26(E)(1)(a) Exception**
   - **310.60(B)(1) Exception**

Where an exception to a rule exists, the exception will immediately follow the main rule to which it applies. An exception is used in the *NEC* only where necessary. When possible, the Technical Committees (*Code*-Making Panels) use positive language within a given section instead of an exception. When an exception is made to a section which contains list items, the exception will clearly indicate the items within the list to which it applies. Where exceptions are used, they truly are an exception to the rule they follow.

> **Fact**
>
> Informative Annexes provide the *NEC* user with additional information, such as a cross-reference to applicable Product Standards in Informative Annex A.

ARTICLE 300
Wiring Methods

I. General

300.12 Mechanical Continuity — Raceways and Cables. Metal or nonmetallic raceways, cable armors, and cable sheaths shall be continuous between cabinets, boxes, fittings, or other enclosures or outlets.

Exception No. 1:   Short sections of raceways used to provide support or protection of cable assemblies from physical damage shall not be required to be mechanically continuous.

Exception No. 2:   Raceways and cables installed into the bottom of open bottom equipment, such as switchboards, motor control centers, and floor or pad-mounted transformers, shall not be required to be mechanically secured to the equipment.

ARTICLE 250
Grounding and Bonding

III. Grounding Electrode System and Grounding Electrode Conductor

250.68 Grounding Electrode Conductor and Bonding Jumper Connection to Grounding Electrodes.

(A) Accessibility. All mechanical elements used to terminate a grounding electrode conductor or bonding jumper to a grounding electrode shall be accessible.

Exception No. 1:   An encased or buried connection to a concrete-encased, driven, or buried grounding electrode shall not be required to be accessible.

Exception No. 2:   Exothermic or irreversible compression connections used at terminations, together with the mechanical means used to attach such terminations to fireproofed structural metal whether or not the mechanical means is reversible, shall not be required to be accessible.

The rule stated in **Section 300.12** requires that all raceways, cable armors, and cable sheaths be continuous between boxes, fittings, or other enclosures or outlets. **Exception No. 1** allows short sections of raceways only (not cable armors and cable sheaths) to provide support and/or protection of cable assemblies. For example, this exception would permit a short section of a raceway, such as electrical metallic tubing (EMT) or rigid metal conduit (RMC), to protect and/or support a cable assembly for a short distance where physical damage could occur to the cable assembly. Note that while the *NEC* does not state a minimum length, the term "short section" infers a piece of a raceway. **Exception No. 2** allows for the floor stub up raceway and cable penetrations into equipment not be terminated to the enclosure.

The rule stated in **250.68(A)** requires that all connections of grounding electrode conductors to grounding electrodes be accessible, with two exceptions. **Exception No. 1** allows for connections encased in concrete or buried being inaccessible. **Exception No. 2** allows for exothermic or irreversible connections to structural steel to be covered in fireproofing materials.

Informational Notes

Informational notes are explanatory material and are not an enforceable part of the *NEC* as required in **90.5(C)**. As such, these notes are variable and include, but are not limited to, the following types:
1. Informational
2. Referential (referencing other sections

Fact

Where an exception is used in the *NEC*, it will immediately follow the rule to which it applies. Exceptions include two types:
1. **Mandatory** where "shall" and "shall not" direct the action
2. **Permissive** where "shall be permitted" states the action is acceptable.

or areas of the *NEC* or other codes)
3. Design
4. Suggestions
5. Examples

Informational notes are used when:
1. Informational notes provide basic information to aid the *NEC* user. Examples of this type of note include the following:
   - **90.5(C) Explanatory Material**
   - **Article 100 Definition of "Listed" Informational Note**
   - **110.11 Informational Note No. 1 and 2**
2. Informational notes provide example(s) of where or how the rule(s) would apply. For example:
   - **250.20 Informational Note**
   - **250.96(B) Informational Note**
3. Informational notes provide reference to other sections within the *NEC* to further explain the requirement and to aid the user in its proper application. For example:
   - **90.7 Informational Note No. 1, 2 and 3**
   - **250.30(A) Informational Note**
   - **314.15 Informational Note No. 1 and 2**
4. Informational notes provide reference to other codes or standards to further explain the requirement, inform of building code requirements, and to aid the user in its proper application. For example:
   - **110.16 Informational Note No. 1 and 2**
   - **210.52(5) Informational Note**
   - **300.21 Informational Note**
5. Informational notes provide a suggestion for adequate performance and/or proper design. For example:
   - **210.4(A) Informational Note**
   - **210.19(A)(1)Exception Informational Note No. 4**

- **215.2(A)(4) Informational Note No. 1**
- **220.61(C)(2) Informational Note No. 1 and 2**
- **240.85 Informational Note**

The following is an example of an informational note.

> **250.20 Alternating-Current Systems to Be Grounded.** Alternating-current systems shall be grounded as provided for in 250.20(A), (B), (C), or (D). Other systems shall be permitted to be grounded. If such systems are grounded, they shall comply with the applicable provisions of this article.
>
> Informational Note: An example of a system permitted to be grounded is a corner-grounded delta transformer connection. See 250.26(4) for conductor to be grounded.

Material for informational notes is included in the *NEC* only where the Code-Making Panel believes that the information is necessary for proper application of the rule(s). In most cases, the text of an informational note is indispensable information for the user and for proper application of the *NEC*. Always read the informational notes. For example, **Article 100 Definitions** includes the following Informational Note for the definition of AJH:

> Informational Note: The phrase "authority having jurisdiction," or its acronym AHJ, is used in NFPA documents in a broad manner, since jurisdictions and approval agencies vary, as do their responsibilities. Where public safety is primary, the authority having jurisdiction may be a federal, state, local, or other regional department or individual such as a fire chief; fire marshal; chief of a fire prevention bureau, labor department, or health department; building official; electrical inspector; or others having statutory authority...

MANDATORY RULES, PERMISSIVE RULES, AND EXPLANATORY MATERIAL

The *NEC* follows specific rules which clearly illustrate whether requirements are mandatory or permissive. These rules also identify text which is explanatory in nature and provided to aid the *Code* user. The *NEC*, in **Article 90 Introduction**, provides the method used to determine the applicability of all text. This article is not subdivided into parts. **Section 90.5** clearly defines the application of all rules and material in the *NEC* as follows:

90.5 Mandatory Rules, Permissive Rules, and Explanatory Material.

(A) Mandatory Rules. Mandatory rules of this Code are those that identify actions that are specifically required or prohibited and are characterized by the use of the terms shall or shall not.

(B) Permissive Rules. Permissive rules of this Code are those that identify actions that are allowed but not required, are normally used to describe options or alternative methods, and are characterized by the use of the terms shall be permitted or shall not be required.

(C) Explanatory Material. Explanatory material, such as references to other standards, references to related sections of this Code, or information related to a Code rule, is included in the Code in the form of informational notes. Such notes are informational only and are not enforceable as requirements of this Code.

Brackets containing section references to another NFPA document are for informational purposes only and are provided as a guide to indicate the source of the extracted text. These bracketed references immediately follow the extracted text.

Informational Note: The format and language used in this Code follows guidelines established by NFPA and published in the NEC Style Manual. Copies of this manual can be obtained from NFPA

(D) Informative Annexes. Nonmandatory information relative to the use of the NEC is provided in informative annexes. Informative annexes are not part of the enforceable requirements of the NEC, but are included for information purposes only.

Mandatory Language

As stated in **90.5(A)**, a rule is considered mandatory by the use of the terms shall or shall not. Where either term is used in the *NEC*, the rule is mandatory unless an exception follows or the rule exists in Chapters 1 through 4 and is supplemented or modified in Chapter 5, 6, or 7. The following is an example of mandatory text:

230.6 Conductors Considered Outside the Building. Conductors shall be considered outside of a building or other structure under any of the following conditions:

(1) Where installed under not less than 50 mm (2 in.) of concrete beneath a building or other structure

(2) Where installed within a building or other structure in a raceway that is encased in concrete or brick not less than 50 mm (2 in.) thick

(3) Where installed in any vault that meets the construction requirements of Article 450, Part III

(4) Where installed in conduit and under not less than 450 mm (18 in.) of earth beneath a building or other structure

(5) Where installed in overhead service masts on the outside surface of the building traveling through the eave of that building to meet the requirements of 230.24

The preceding rule in **Article 230 Services**, in **Part I. General** requires that service conductors (conductors meeting the definition of "service conductors" in **Article 100**) that meet any of the conditions in list items 1 through 5 be considered as being outside of the building.

Permissive Text

As stated in **90.5(B)**, a rule is considered permissive by the use of the terms "shall be permitted" or "shall not be required." Where these terms are used in the *NEC*, the rule is permissive unless an exception follows the rule or unless the rule exists in Chapters 1 through 4 and is supplemented or modified in Chapters 5 through 7. See the following example of permissive text:

> **250.20 Alternating-Current Systems to Be Grounded.** Alternating-current systems shall be grounded as provided for in 250.20(A), (B), (C), or (D). Other systems shall be permitted to be grounded. If such systems are grounded, they shall comply with the applicable provisions of this article.
>
> Informational Note: An example of a system permitted to be grounded is a corner-grounded delta transformer connection. See 250.26(4) for conductor to be grounded.

This requirement, in **Article 250 Grounding** in **Part II. System Grounding**, mandates the following:

1. Alternating-current systems shall be grounded as provided for in 250.20(A), (B), (C), or (D), and
2. Other systems shall be permitted to be grounded. If such systems are grounded, they shall comply with the applicable provisions of this article.

The second requirement of this section specifically permits systems other than those illustrated in **250.20(A)**, **(B)**, **(C)**, or **(D)** to be grounded by the use of the term "shall be permitted." This wording is, in essence, a mandated permission. The alternative would be for the *NEC* to use the term "may" to illustrate permissiveness. However, "may" is considered unenforceable and is not used because it could be taken to mean that the inspector "may" permit other systems to be grounded or he/she "may not." The *NEC Style Manual* lists terms to be avoided, such as "may," which are vague and/or unenforceable in an effort to make the *Code* as clear and usable as possible. **See Figure 3-4.**

Other vague terms the *NEC* avoids include: acceptable, adequate, appropriate, desirable, familiar, frequent, generally, likely, may, most, near, practices, prefer, proper, reasonable, satisfactory, significant, sufficient, suitable, and workmanlike.

Units of Measurement

The *NEC* uses both the SI system (metric units) and inch-pound units. The SI system is always shown first, followed by the inch-pound in parentheses. For example:

> **334.30 Securing and Supporting.** Nonmetallic-sheathed cable shall be supported and secured by staples, cable ties, straps, hangers, or similar fittings designed and installed so as not to damage the cable at intervals not exceeding 1.4 m (4 1/2 ft) and within 300 mm (12 in.) of every outlet box, junction box, cabinet, or fitting. Flat cables shall not be stapled on edge.

Sections of cable protected from physical damage by raceway shall not be required to be secured within the raceway.

Section 334.30, in **Article 334 Nonmetallic-Sheathed Cable: Types NM, NMC, and NMS**, in **Part II. Installation**, requires that Type NM cable be secured and supported. Note that the SI system is always listed first and the inch-pound units second and always in parentheses. Note also that **90.9(D)** specifically permits the use of either the SI system or the inch-pound system.

> *Fact*
>
> The SI system (metric units) is always shown first in the *NEC* with the inch-pound units shown second in parentheses because this *Code* is used internationally.

Figure 3-4	*NEC* Text Rules	
TYPE OF TEXT	**LISTED TERM(S)**	
Mandatory Text	"Shall" "Shall not"	
Permissive Text	"Shall be permitted"	"Shall not be required"
Explanatory Text	Informational Notes	Informative Annex

Figure 3-4. Text Language in the NEC includes mandatory, permissive, and explanatory rules.

Extract Material

The *National Electrical Code*, also known as *NFPA 70*, is one of many documents published by the National Fire Protection Association. Where another NFPA document has primary jurisdiction over material to be included in the *NEC*, the material is extracted into the *NEC*. An example of another NFPA document which would have primary jurisdiction over material addressed by the *NEC* is *NFPA 20-2010, Standard for the Installation of Stationary Pumps for Fire Protection*. When such extraction occurs, the document in which the extract material exists is identified at the beginning of the article. An Informational Note immediately follows the title of the article to inform the *Code* user of the presence of extract material. Rules within the article which are extracted are followed with the title of the referenced NFPA document and sections in brackets. Referencing sections of another NFPA document is for informational purposes only, as stated in **90.5(C)**. An example of extract material in the *NEC* can be seen in **Article 695**:

> **ARTICLE 695**
> **Fire Pumps**
>
> **695.1 Scope.**
>
> Informational Note: Text that is followed by a reference in brackets has been extracted from NFPA 20-2007, Standard for the Installation of Stationary Pumps for Fire Protection. Only editorial changes were made to the extracted text to make it consistent with this Code.

The informational note following the title of **Article 695** informs the user that it contains extract material from *NFPA 20-2010, Standard for the Installation of Stationary Pumps for Fire Protection*. The informational note further explains that where this material is located, it is identified with a reference in brackets at the end of the rule. **See Figure 3-5.** This extracted material can only be changed editorially to fit the style of the *NEC*. The following section from **Article 695** contains extracted material as follows:

> **ARTICLE 695**
> **Fire Pumps**
> **695.10 Listed Equipment.** Diesel engine fire pump controllers, electric fire pump controllers, electric motors, fire pump power transfer switches, foam pump controllers, and limited service controllers shall be listed for fire pump service. [20: 9.5.1.1, 10.1.2.1, 12.1.3.1]

The bracketed text located at the end of **Section 695.10** identifies the sections and title of the document from which the extract material originated.

Figure 3-5 — Bracketed Text Throughout the *NEC*

ARTICLE 695
Fire Pumps

695.1 Scope.

Informational Note: Text that is followed by a reference in brackets has been extracted from NFPA 20-2010, *Standard for the Installation of Stationary Pumps for Fire Protection*. Only editorial changes were made to the extracted text to make it consistent with this *Code*.

(A) Covered. This article covers the installation of the following:

(1) Electric power sources and interconnecting circuits

(2) Switching and control equipment dedicated to fire pump drivers

(B) Not Covered. This article does not cover the following:

Reprinted with permission from NFPA 70-2011, *National Electrical Code®*, Copyright© 2010, National Fire Protection Association, Quincy, MA 02169. This reprinted material is not the complete and official position of the NFPA on the referenced subject, which is represented only by the standard in its entirety.

Figure 3-5. Throughout the NEC, references are made to other NFPA documents which have primary jurisdiction over the material.

Cross-Reference Tables

The *NEC* provides six cross-reference tables where multiple modifications or supplemental requirements related to the scope of an article are located elsewhere in the *NEC*. **See Figure 3-6.**

Table 225.3 is a cross-reference table for outside branch circuit and feeders installations for various locations. **See Figure 3-7.** Some of these locations include:

- Community television/radio systems
- Electrically driven irrigation machines
- Electric Signs
- Fixed electric deicing and snow-melting equipment
- Floating buildings
- Hazardous locations such as fueling stations
- Marinas and boatyards
- Solar photovoltaic systems
- Swimming pools, fountains, etc.

Table 225.3 cross-references requirements for various outside branch circuit and feeder installations.

Figure 3-6	Cross-Reference Tables	
TYPE OF TEXT	**LISTED TERMS**
ARTICLE 210 Branch Circuits **Table 210.2 Specific-Purpose Branch Circuits**	Provides cross-references to aid the *Code* user in 33 other locations.
ARTICLE 220 Branch-Circuit Feeder and Service Calculations **Table 220.3 Additional Load Calculation References**	Provides cross-references to aid the *Code* user in 30 other locations.
ARTICLE 225 Outside Branch Circuit and Feeders **Table 225.3 Other Articles**	Provides cross-references to aid the *Code* user in 25 other articles.
ARTICLE 240 Overcurrent Protection **Table 240.3 Other Articles**	Provides cross-references to aid the *Code* user in 35 other articles.
ARTICLE 250 Grounding and Bonding **Table 250.3 Additional Grounding Requirements**	Provides cross-references to aid the *Code* user in over 90 other locations.
ARTICLE 430 Motors, Motor Circuits, and Controllers **Table 430.5 Other Articles**	Provides cross-references to aid the *Code* user in 23 other locations.

Figure 3-6. There are six cross-reference tables in the NEC which list multiple changes to the scope of an article.

Table 225.3 Other Articles

Equipment/Conductors	Article
Branch circuits	210
Class 1, Class 2, and Class 3 remote-control, signaling, and power-limited circuits	725
Communications circuits	800
Community antenna television and radio distribution systems	820
Conductors for general wiring	310
Electrically driven or controlled irrigation machines	675
Electric signs and outline lighting	600
Feeders	215
Fire alarm systems	760
Fixed outdoor electric deicing and snow-melting equipment	426
Floating buildings	553
Grounding and bonding	250
Hazardous (classified) locations	500
Hazardous (classified) locations— specific	510
Marinas and boatyards	555
Messenger-supported wiring	396
Mobile homes, manufactured homes, and mobile home parks	550
Open wiring on insulators	398
Over 600 volts, general	490
Overcurrent protection	240
Radio and television equipment	810
Services	230
Solar photovoltaic systems	690
Swimming pools, fountains, and similar installations	680
Use and identification of grounded conductors	200

725.141 Installation of Circuit Conductors Extending Beyond One Building. Where Class 2 or Class 3 circuit conductors extend beyond one building and are run so as to be subject to accidental contact with electric light or power conductors operating over 300 volts to ground, or are exposed to lightning on interbuilding circuits on the same premises, the requirements of the following shall also apply:

(1) Sections 800.44, 800.50, 800.53, 800.93, 800.100, 800.170(A), and 800.170(B) for other than coaxial conductors

(2) Sections 820.44, 820.93, and 820.100 for coaxial conductors

725.143 Support of Conductors. Class 2 or Class 3 circuit conductors shall not be strapped, taped, or attached by any means to the exterior of any conduit or other raceway as a means of support. These conductors shall be permitted to be installed as permitted by 300.11(B)(2).

725.154 Applications of Listed Class 2, Class 3, and PLTC Cables. Class 2, Class 3, and PLTC cables shall comply with any of the requirements described in 725.154(A) through (I).

(A) Plenums. Cables installed in ducts, plenums, and other spaces used for environmental air shall be Type CL2P or CL3P. Listed wires and cables installed in compliance with 300.22 shall be permitted. Listed plenum signaling race-

VII. Service Equipment — Overcurrent Protection

230.90 Where Required. Each ungrounded service conductor shall have overload protection.

(A) Ungrounded Conductor. Such protection shall be provided by an overcurrent device in series with each ungrounded service conductor that has a rating or setting not higher than the allowable ampacity of the conductor. A set of fuses shall be considered all the fuses required to protect all the ungrounded conductors of a circuit. Single-pole circuit breakers, grouped in accordance with 230.71(B), shall be considered as one protective device.

Exception No. 1: For motor-starting currents, ratings that comply with 430.52, 430.62, and 430.63 shall be permitted.

Exception No. 2: Fuses and circuit breakers with a rating or setting that complies with 240.4(B) or (C) and 240.6 shall be permitted.

Figure 3-7. A cross-reference table such as Table 225.3 (Article 225 Outside Branch Circuits and Feeders) can reference several articles and sections.

Outlines, Diagrams, and Drawings

The *NEC* does not contain pictures to aid the *Code* user in the application of installation requirements. **90.1(C)** clearly states that the *NEC* is not intended as a design manual or as an instruction manual for untrained persons. However, the *NEC* does include outlines, diagrams, and drawings which describe the application of an article or section or provide basic information in ladder-type diagrams to aid the *Code* user. These informational outlines are provided as follows:

90.3 Code Arrangement.

Figure 90.3 Code Arrangement.

The illustration provided with this section is designed to clearly outline the arrangement and application of rules contained in the *NEC*.

210.52 Dwelling Unit Receptacle Outlets.

Figure 210.52(C)(1) Determination of Area Behind a Range, or Counter-Mounted Cooking Unit or Sink.

This drawing is provided to aid the user of this code in applying the receptacle outlet requirement of **210.52(C)(1)** near a sink, counter-mounted cooking unit or range.

220.1 Scope.

Figure 220.1 Branch Circuit, Feeder, and Service Calculation Methods.

This illustration is included to aid the *Code* user in understanding the permitted calculation methods of **Article 220**.

230.1 Scope.

Figure 230.1 Services.

The diagram provided is a useful outline of how to apply *NEC* rules for services from the service point to the premises wiring. **See Figure 3-8.**

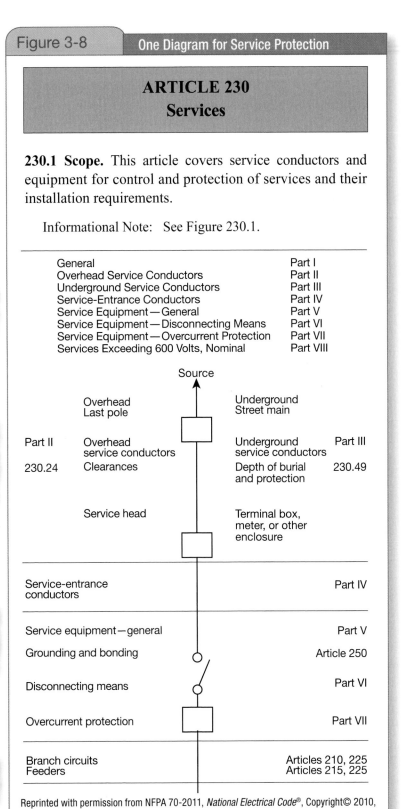

Figure 3-8 | **One Diagram for Service Protection**

ARTICLE 230
Services

230.1 Scope. This article covers service conductors and equipment for control and protection of services and their installation requirements.

Informational Note: See Figure 230.1.

General	Part I
Overhead Service Conductors	Part II
Underground Service Conductors	Part III
Service-Entrance Conductors	Part IV
Service Equipment—General	Part V
Service Equipment—Disconnecting Means	Part VI
Service Equipment—Overcurrent Protection	Part VII
Services Exceeding 600 Volts, Nominal	Part VIII

Figure 3-8. The NEC includes many figures to aid the Code user.

250.1 Scope.

Figure 250.1 Grounding and Bonding.

This illustration details the organization of **Article 250 Grounding and Bonding**.

430.1 Scope.

Figure 430.1 Article 430 Contents.

The single-line diagram provided is a useful outline of how to apply *NEC* rules for motors.

514.3 Classification of Locations.

Figure 514.3 Classified Areas Adjacent to Dispensers as Detailed in Table 514.3(B)(1). [30A: Figure 8.3.1]

This drawing is included to aid the *Code* user in applying **514.3(B)(1)** in classified locations adjacent to dispensers. **See Figure 3-9.**

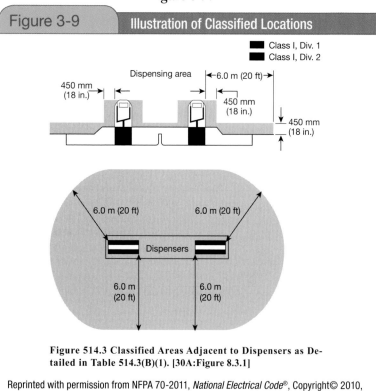

Figure 3-9   **Illustration of Classified Locations**

Figure 514.3 Classified Areas Adjacent to Dispensers as Detailed in Table 514.3(B)(1). [30A:Figure 8.3.1]

Reprinted with permission from NFPA 70-2011, *National Electrical Code®*, Copyright© 2010, National Fire Protection Association, Quincy, MA 02169. This reprinted material is not the complete and official position of the NFPA on the referenced subject, which is represented only by the standard in its entirety.

Figure 3-9. Requirements for Class-1 fueling facility environments are covered in Article 514.

516.3 Classification of Locations.

Figure 516.3(C)(1), (2), (3), and (4)

Figures 516.3(C)(1), (2), (3), and (4) aid the *Code* user in applying the class and division requirements for electrical areas around open spray areas and spray booths. For example, Figure 516.3(C)(1) applies specifically to open spraying. It declares that the actual spray area is a Class I Division I area, whereas all space outside of but within 20 feet horizontally and 10 feet vertically is either defined as Class I, Division 2; Class, Zone 2; or Class II, Division 2. *The definition of Classes, Divisions, and Zones are beyond the scope of this textbook.*

517.30 Essential Electrical Systems for Hospitals.

Informational Note Figure 517.30, No. 1 and 2

These single-line diagrams provide a useful tool for the *Code* user to understand and apply the rules of **Article 517** for essential electrical systems for hospitals.

517.41 Essential Electrical Systems.

Informational Note Figure 517.41, No. 1 and 2

These single-line diagrams provide a useful tool for the *Code* user to understand and apply the rules of **Article 517** for nursing homes and limited health care facilities.

550.10(C) Attachment Plug Cap.

Figure 550.10(C) 50-Ampere, 125/250-Volt Receptacle and Attachment Plug Cap Configurations, 3-Pole, 4-Wire, Grounding-Types, Used for Mobile Home Supply Cords and Mobile Home Parks.

551.46(C) Attachment Plugs.

Figure 551.46(C) Configurations for Grounding-Type Receptacles and Attachment Plug Caps Used for Recreational Vehicle Supply Cords and Recreational Vehicle Lots.

552.44(C) Attachment Plugs.

Figure 552.44(C) Attachment Cap and Receptacle Configurations.

These similar drawings depict required configurations for attachment plug caps and grounding-type receptacles for the following: **Article 550 Mobile Homes, Manufactured Homes, and Mobile Home Parks**; **Article 551 Recreational Vehicles and Recreational Vehicle Parks**; and **Article 552 Park Trailers**.

620.2 Control System.
Figure 620.2 Control System.

This flow chart illustrates the elevator control system.

620.13 Feeder and Branch Circuit Conductors.
Figure 620.13 Single Line Diagram.

The single-line diagram provided in **Figure 620.13** is a useful outline of how to apply *NEC* rules for elevators.

680.8 Overhead Conductor Clearances.
Figure 680.8 Clearances From Pool Structures.

The drawing in **Figure 680.8** together with **Table 680.8** provides the *Code* user with a visual explanation of overhead clearances.

690.1 Scope.
Figure 690.1(A) and (B)

The drawings and diagrams of **Figure 690.1(A)** and **Figure 690.1(B)** help the *Code* user identify solar photovoltaic system components as well as common system configurations.

725.121 Power Sources for Class 2 and Class 3 Circuits.
Figure 725.121 Class 2 and Class 3 Circuits.

Figure 725.121 illustrates the relationships between Class 2 or Class 3 power sources, their supply, and the Class 2 or Class 3 circuits.

760.154(D) Fire Alarm Cable Substitutions.
Figure 760.154(D) Cable Substitution Hierarchy.

Figure 760.154(D) illustrates the substitution hierarchy for fire alarm cables.

800.154 Applications of Listed Communications Wires, Cables, and Raceways.
Figure 800.154 Cable Substitution Hierarchy.

This figure helps to describe communicational cable substitution hierarchy.

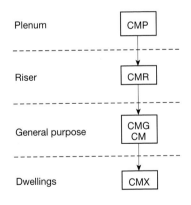

Figure 3-10   Communications Cable Substitutions

Table 800.154(b) Cable Substitutions

Cable Type	Permitted Substitutions
CMR	CMP
CMG, CM	CMP, CMR
CMX	CMP, CMR, CMG, CM

Plenum — CMP

Riser — CMR

General purpose — CMG CM

Dwellings — CMX

Type CM—Communications cables

[A] Cable A shall be permitted to be used in place of Cable B.

[B]

Figure 800.154 Cable Substitution Hierarchy.

Figure 3-10. Communications cable types (CMP, CMR, CMG/CM, and CMX) are permitted in specific locations as stated in Article 800.

Summary

One of the most basic and crucial steps to become proficient users of the *National Electrical Code* is to gain an in-depth understanding of the structure of the *NEC* text. Its structure is governed by the *NEC Style Manual* through a detailed outline form followed consistently throughout the *Code*. An understanding of this outline form is essential to the proper application of *NEC* requirements. The structure of the *NEC* is summarized as follows:

- The *NEC* is divided into 10 major parts: the introduction and nine chapters.
- Chapters, the major subdivisions of the *NEC*, are charged with broad scopes. The scope of each chapter is subdivided logically into separate articles to address each chapter scope.
- Articles are major subdivisions of chapters which cover a specific topic within the scope of the chapter. When an article is of large size or when usability is an issue, it is subdivided into separate parts, each dedicated to a logical separation of requirements within the given article.
- Parts are major subdivisions of articles which logically separate information for ease of use and proper application. Parts are then broken down into separate sections and individually titled to address the scope of the individual part.
- Sections may be logically subdivided into three levels. Sections may also contain list items, exceptions, and informational notes.
- The structure of the *NEC* is in a progressive ladder-type format, which when applied is as follows:
  - A rule which exists in a third-level subdivision applies only under:
  - The rule in the second-level subdivision, which is limited to:
  - The rule in the first-level subdivision, which applies only under:
  - The section in which it exists, which applies only in:
  - The part it is arranged in, which applies only in:
  - The article in which it exists, which is limited to:
  - The chapter in which it is located.
- List items are used in sections, subdivisions, or exceptions where necessary.
- Exceptions are used only when absolutely necessary and are always italicized.
- Informational notes are informational only and are not mandatory.
- Tables in Chapter 9 are applicable only where referenced elsewhere in the *NEC*.
- Informative Annexes are informational only and are not mandatory.
- Mandatory language in the *NEC* consists of the use of "shall" and "shall not."
- Permissive language in the *NEC* consists of the use of "shall be permitted" and "shall not be required."
- Explanatory text exists in the form of informational notes.
- Cross-references, outlines, drawings, and diagrams are included to aid the user in the proper application of the *NEC*.

An in-depth understanding of the structure or outline form of the *NEC* is necessary to properly apply the requirements of the *Code*. Understanding this structure or outline form of the *NEC* is one of the cornerstones of the *Codeology* method.

1. Where multiple modifications or supplemental requirements are related to the scope of an article located elsewhere in the *NEC*, an article will contain a(n) __?__ table to aid the *Code* user.
   a. calculations
   b. contents
   c. conduit fill
   d. cross-reference

2. Which one of the following is an example of permissive language in the *NEC*?
   a. "May be permitted"
   b. "If the installer desires"
   c. "Shall be permitted"
   d. "May if cost is an issue"

3. Exceptions are used in the *NEC* __?__ .
   a. only when necessary
   b. to confuse the *Code* user
   c. to justify shortcuts
   d. to allow alternate methods

4. Informational notes are used in the *NEC* to aid the *Code* user and are __?__ .
   a. informational only
   b. explanatory material
   c. designed to aid the *Code* user
   d. All of the above

5. When an article is subdivided into logical separations, these subdivisions are called __?__ .
   a. sections
   b. parts
   c. subdivisions
   d. annexes

6. Parts are subdivided into logical separations called __?__ .
   a. sections
   b. parts
   c. subdivisions
   d. informative annexes

7. Which one of the following is an example of mandatory language in the *NEC*?
   a. "Shall not"
   b. "You really should not"
   c. "May not"
   d. "Must not"

8. Which of the following types of informational notes are designed to aid the user of the *NEC*?
   a. Informational
   b. Reference
   c. Design suggestions and/or examples
   d. All of the above

Article 100: The Language of the *NEC*

Throughout this text, basic principles will be repeated time and time again until they become part of the natural thought process when working with the *NEC*. When reviewing the *NEC*, or any other type of installation code, it is absolutely necessary to understand the language of the *Code*. The language of the *NEC* is primarily outlined by definitions in **Article 100** and occasionally by definitions within the second section of individual articles.

Objectives

» Recognize that the *NEC* **Article 100** defines terms to aid the user in the understanding of the language used throughout the *Code*.

» Determine that the *NEC* defines terms in **Article 100** only when the definition is essential to proper application or when the terms are used in two or more articles.

» Understand that the *NEC* does not define common terms or technical terms that are commonly understood.

» Recognize that definitions do not contain requirements.

» Locate additional definitions not found in **Article 100** in section two of the article where they appear when the terms need to be defined, but only exist in a single article.

Chapter 4

Table of Contents

NEC LANGUAGE: ARTICLE 100

Article 100 contains only those definitions which are essential to the proper application of this *Code*. The intent is not to include commonly defined general terms or commonly defined technical terms from related codes and standards. In general, only those terms which are used in two or more articles are defined in **Article 100**. Other definitions are included in the article in which they are used, but may be referenced in **Article 100**.

Part **I. General** of **Article 100** contains definitions intended to apply wherever the terms are used throughout the *Code*. Part **II. Over 600 Volts, Nominal** contains definitions applicable only to the parts of articles specifically covering installations and equipment operating at more than 600 volts, nominal.

NEC LANGUAGE: OVERVIEW

Users of the *National Electrical Code* must learn the definitions of terms utilized within the document. The *NEC* might seem to be written in a different language; therefore, **Article 100 Definitions** will provide a framework of the terms. Terms which are defined must follow the *NEC Style Manual*. As stated above, the *Code* does not attempt to define commonly used general terms or commonly used technical terms from related codes and standards. Only those terms which are essential to the proper application of the *NEC* are defined. Unless a term is used in two or more articles, it will not appear in **Article 100** and instead be defined in the article in which it is used. Definitions are presented as follows:

- Definitions are in alphabetical order.
- Definitions do not contain the term being defined.
- Definitions do not contain requirements or recommendations.
- Defined terms which appear in two or more articles are listed in Article 100.
- Defined terms which appear in a single

article are listed in the second section (for example, 680.2) of that article.

All users of the *NEC* must be familiar with definitions in order to properly interpret and apply each requirement. Moreover, the *NEC* definition for what seems to be a "common term" might differ from a standard dictionary explanation, therefore seriously impacting the implementation of an electrical installation. For example:

Q. *A metal-sheathed cable assembly such as type AC cable is installed above a lay-in type acoustical ceiling. The cable is installed in accordance with the NEC, and the ceiling tiles have been replaced. The AC cable, now hidden behind the lay-in tiles of the drop ceiling, can no longer be seen. Is the AC cable considered to be "exposed"?*

A. *Yes, the AC cable above the lay-in type acoustical ceiling is installed in accordance with the NEC definition of exposed. According to the standard dictionary, the term "exposed" would normally be considered to mean that the item being discussed could be seen. However, the Code defines the term for two different scenarios:*

- *Exposed (as applied to live parts)*
- *Exposed (as applied to wiring methods)*

Because this question is about type AC cable, which is a wiring method, the second definition of exposed would apply:

Article 100 Definitions.
Exposed (as applied to wiring methods). On or attached to the surface or behind panels designed to allow access.

Note that this definition would apply to the type AC cable installed above a lay-in type acoustical ceiling because it is installed behind panels (lay-in ceiling tile) designed to allow access. **See Figure 4-1.** This definition would also apply to receptacles and junction boxes under a raised computer floor. **See Figure 4-2.** The term "exposed" is used more than 300 times in the *NEC*. All users of the *Code* must be familiar with the most common terms used throughout the document.

Many articles define terms used only within the same article. The *NEC Style Manual* requires that such term be listed within the second section of that article, so that the *NEC* user can see early on the terms defined for proper application of the requirements which follow. The first section is always the scope of an article, and the second is a list of definitions of terms used within that article.

For example, in **Article 517 Health Care Facilities** the second section of the article, **Section 517.2 Definitions**, lists 40 terms and definitions. Remember, a term must be defined in **Article 100** if it is used in two or more articles. Article 100 contains more than 150 definitions.

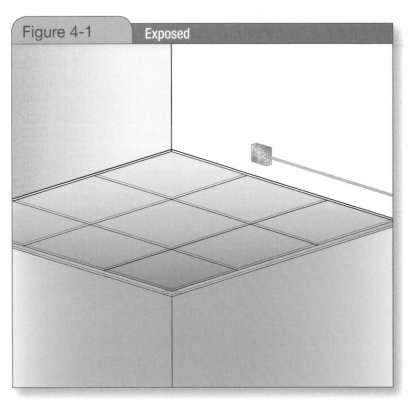

Figure 4-1. A box above a lay-in ceiling is defined as exposed (as applied to wiring methods) under Article 100 Definitions.

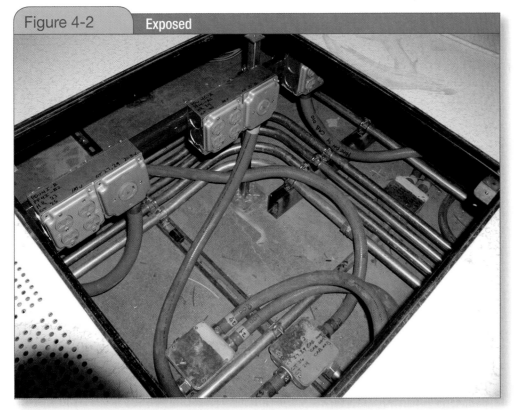

For additional information, visit qr.njatcdb.org Item #1063

Figure 4-2. Receptacles and junction boxes below a raised computer floor are defined as exposed (as applied to wiring methods) under Article 100 Definitions.

NEC LANGUAGE: DEFINITIONS

The following **Article 100** definitions are examples of the commonly used terms in the *NEC* which the *Code* user must clearly understand for proper application of its requirements.

> **Accessible, Readily (Readily Accessible).** Capable of being reached quickly for operation, renewal, or inspections without requiring those to whom ready access is requisite to climb over or remove obstacles or to resort to portable ladders, and so forth.

The term "readily accessible" is used more than 70 times in the *NEC*. Other terms related to accessibility and defined in **Article 100** are accessible (as applied to equipment), accessible (as applied to wiring methods), and concealed. **See Figure 4-3.**

If the disconnect switch shown in **Figure 4-3** was located behind a stack of boxes, it would be considered accessible, not readily accessible. Also, if the discon-

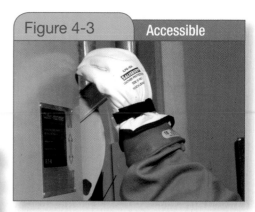

Figure 4-3 | Accessible

Figure 4-3. A person operating equipment which is readily accessible.

nect switch had a padlock installed, again it would only be accessible.

> **Ampacity.** The maximum current, in amperes, that a conductor can carry continuously under the conditions of use without exceeding its temperature rating.

The term "ampacity" is used more than 400 times in the *NEC* and is derived from the combination of the terms ampere and capacity. **See Figure 4-4.**

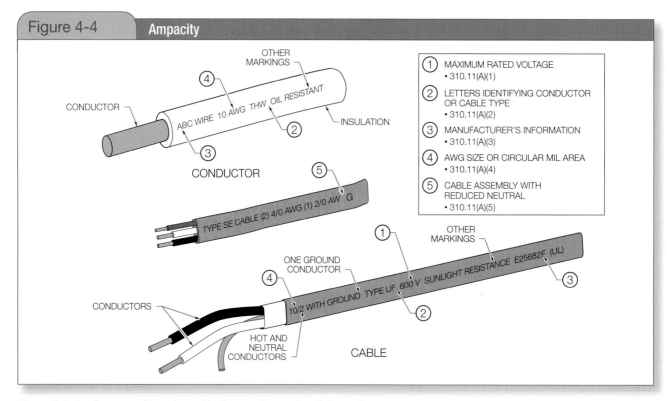

Figure 4-4 | Ampacity

CONDUCTOR — OTHER MARKINGS — INSULATION

(4) (2) (3)

ABC WIRE 10 AWG THW OIL RESISTANT

CONDUCTOR

(5)

TYPE SE CABLE (2) 4/0 AWG (1) 2/0 AWG

(1) MAXIMUM RATED VOLTAGE
  • 310.11(A)(1)
(2) LETTERS IDENTIFYING CONDUCTOR OR CABLE TYPE
  • 310.11(A)(2)
(3) MANUFACTURER'S INFORMATION
  • 310.11(A)(3)
(4) AWG SIZE OR CIRCULAR MIL AREA
  • 310.11(A)(4)
(5) CABLE ASSEMBLY WITH REDUCED NEUTRAL
  • 310.11(A)(5)

ONE GROUND CONDUCTOR

CONDUCTORS

(4) (1) OTHER MARKINGS

10/2 WITH GROUND TYPE UF 600 V SUNLIGHT RESISTANCE E25682F (UL)

(2) (3)

HOT AND NEUTRAL CONDUCTORS

CABLE

Figure 4-4. Conductor markings do not list the actual ampacity of the conductor.

Approved. Acceptable to the authority having jurisdiction.

"Approved" is used more than 300 times in the *NEC*. It is essential that the *Code* user know that when this term is used, a compliancy is determined by the "authority having jurisdiction" (AHJ) and not a third-party listing organization such as UL.

The term "authority having jurisdiction" is also defined in **Article 100**. The informational note following the definition states that the AHJ may be federal, state, local, or other regional departments or individuals such as a fire chief, fire marshal, labor department, health department, electrical inspector, or others having statutory authority. The note further states that for insurance purposes, an insurance inspection department (i.e. FM), rating bureau, or other insurance company representatives may be the AHJ.

Branch Circuit. The circuit conductors between the final overcurrent device protecting the circuit and the outlet(s).

Building. A structure that stands alone or that is cut off from adjoining structures by fire walls with all openings therein protected by approved fire doors.

The *NEC* contains specific requirements and limitations for buildings, so it is essential to understand that a row of ten strip stores, for example, may be recognized in the *NEC* as 10 separate buildings; provided they are separated by fire walls. **See Figure 4-5.**

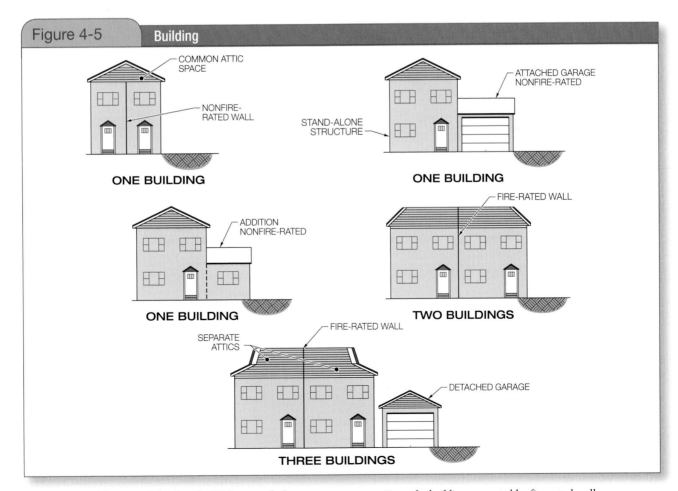

Figure 4-5 | Building

COMMON ATTIC SPACE

NONFIRE-RATED WALL

ONE BUILDING

ATTACHED GARAGE NONFIRE-RATED

STAND-ALONE STRUCTURE

ONE BUILDING

ADDITION NONFIRE-RATED

ONE BUILDING

FIRE-RATED WALL

TWO BUILDINGS

SEPARATE ATTICS

FIRE-RATED WALL

DETACHED GARAGE

THREE BUILDINGS

Figure 4-5. Buildings are defined in the NEC as stand-alone structures or portions of a building separated by fire-rated walls.

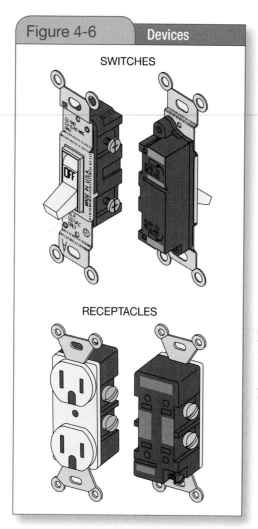

Figure 4-6. Devices

SWITCHES

RECEPTACLES

Figure 4-6. Switches and receptacles are defined as devices in Article 100.

Device. A unit of an electrical system that carries or controls electric energy as its principal function.

The term device is used in the *NEC* more than 500 times. Types of devices would include but not be limited to, receptacles and switches. **See Figure 4-6.**

Exposed (as applied to wiring methods). On or attached to the surface or behind panels designed to allow access.

The term "exposed" as used in the *NEC* is very different from the standard use of

this term. Other related definitions in **Article 100** include exposed (as applied to live parts) and Concealed.

Feeder. All circuit conductors between the service equipment, the source of a separately derived system, or other power supply source and the final branch-circuit overcurrent device.

For proper application of the *NEC*, the *Code* user must be able to determine the proper *Code* term for all current-carrying conductors. The four types of current-carrying conductors are branch circuit, feeder, service, and tap conductors.

Fitting. An accessory such as a locknut, bushing, or other part of a wiring system that is intended primarily to perform a mechanical rather than an electrical function.

Fittings are used to terminate conduit and other raceways to boxes and enclosures. **See Figure 4-7.**

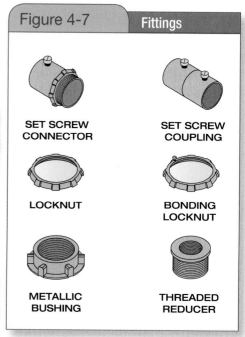

Figure 4-7. Fittings

SET SCREW CONNECTOR

SET SCREW COUPLING

LOCKNUT

BONDING LOCKNUT

METALLIC BUSHING

THREADED REDUCER

Figure 4-7. Fittings only perform a mechanical function such as termination of conduit to an enclosure.

In Sight From (Within Sight From, Within Sight). Where this Code specifies that one equipment shall be "in sight from," "within sight from," or "within sight of," and so forth, another equipment, the specified equipment is to be visible and not more than 15 m (50 ft) distant from the other.

When the *NEC* requires that equipment, such as a motor and an associated disconnecting means be "within sight of" each other, the requirement is that the equipment be visible and not more than 50 feet apart.

Outlet. A point on the wiring system at which current is taken to supply utilization equipment.

The term "outlet" would include a receptacle, a ceiling-mounted box for a lighting fixture, and the point at which equipment is hard-wired. **See Figure 4-8.**

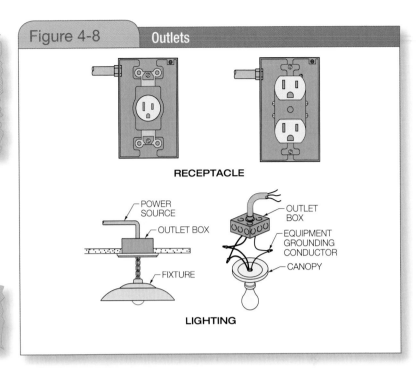

Figure 4-8 | Outlets

RECEPTACLE

POWER SOURCE
OUTLET BOX
FIXTURE
OUTLET BOX
EQUIPMENT GROUNDING CONDUCTOR
CANOPY

LIGHTING

Figure 4-8. Outlets are not only receptacles, but include lighting outlets in the ceiling.

Overcurrent. Any current in excess of the rated current of equipment or the ampacity of a conductor. It may result from overload, short circuit, or ground fault.

An overcurrent includes the following:
• Overload: above the normal full-load rating. For example, 22 amps flowing on a 20-amp branch circuit is an overload.
• Short Circuit: when the current does not flow through its normal path (it takes a shortcut), but continues on circuit conductors.

Ground Fault. An unintentional, electrically conducting connection between an ungrounded conductor of an electrical circuit and the normally non–current-carrying conductors, metallic enclosures, metallic raceways, metallic equipment, or earth.

• Ground Fault: a type of short circuit in which current flows outside of the circuit conductors and returns to the source through grounded equipment or an equipment grounding conductor.

Qualified Person. One who has skills and knowledge related to the construction and operation of the electrical equipment and installations and has received safety training to recognize and avoid the hazards involved.

The term "qualified person" is used more than 100 times in the *NEC*. It is extremely important that all *Code* users understand that a qualified person has specific skills and knowledge as well as safety training.

The informational note refers the user to *NFPA 70E Standard for Electrical Safety in the Workplace* for electrical safety training requirements.

Service Conductors. The conductors from the service point to the service disconnecting means.

Summary

One of the most basic and crucial steps to being proficient in the use of the *National Electrical Code* is an in-depth understanding of the language of the *NEC*. Chapter 1 of this text discussed the development procedures of the *NEC*; Chapter 2 explained the Table of Contents and Arrangement of the *NEC*; and Chapter 3 introduced the Organization of the *NEC*. In this chapter, definitions and common terms of *NEC* **Article 100** are listed.

The *NEC* does not include common general terms or common technical terms from related codes and standards. Only those terms which are essential to the proper application of the *NEC* are defined. When a term is essential to the proper application of the *NEC* in two or more articles, its definition is placed in **Article 100**. When a term is defined but used in only a single article, its definition is placed in the second section of that particular article.

For example, the term "substation" is only defined in Outside Branch Circuits and Feeder **Section 225.2** because in **Article 225**, it is essential to the proper application of the *Code*:

Substation: An enclosed assemblage of equipment (e.g., switches, circuit breakers, buses and transformers) under the control of qualified persons, through which electric energy is passes for the purpose of switching or modifying its characteristics.

Though the "substation" term is used in several other articles (see below), it is not considered essential to the proper application within those articles. It is for this reason that "substation" is not defined in **Article 100**. But it is possible that in the future, the *NEC* committees may feel the term has become essential to proper application throughout the *NEC* and will define the term in **Article 100 (Definitions)**.

Areas where the term "Substation" appears in the *NEC*

Introduction
90.2(A)(2)&(4)
Requirements for Electrical Installations
110.31(B)(1), 110.51(A)
Outside Branch Circuits and Feeders
225.2, 225.70(A)(5)Exception to (a)
Overcurrent Protections
240.2
Grounding and Bonding
250.188, 250.191
Equipment, Over 600 Volts, Nominal
490.21(B)(7), 490.30, 490.51
Motion Picture and Television Studios and Similar Locations
530.18(B), 530.19(A), 530.61, 530.62, 530.64

Definitions in **Article 100** do not contain requirements and are listed in alphabetical order. Reading and applying the *NEC* can be as difficult as understanding a foreign language without an in-depth understanding of the definitions within its requirements. Understanding this language and the terms defined in the *NEC*, is one of the cornerstones of the *Codeology* method.

1. The *National Electrical Code* defines only terms which are essential for __?__ of the *NEC*.
    a. spelling terms
    b. changing
    c. printing
    d. proper application

2. A lack of understanding of a term defined in the *NEC* will result in which of the following?
    a. Misapplication of the *NEC*
    b. Confusion
    c. Violations
    d. All of the above

3. Definitions are placed in Article 100 only when the term being defined is used in __?__ or more articles.
    a. four
    b. two
    c. six
    d. three

4. Where a defined term exists in only one article, the term is defined in the __?__ section of the article.
    a. second
    b. largest
    c. most appropriate
    d. first

5. Definitions are not permitted to contain __?__.
    a. the term being defined
    b. a requirement
    c. both a and b
    d. none of the above

6. Definitions are always placed in __?__ order.
    a. chronological
    b. numerical
    c. alphabetical
    d. reverse

7. Definitions in Article 100 apply __?__ in the *NEC*.
    a. globally
    b. sparingly
    c. intermittently
    d. optionally

8. A Type AC cable installed above a lay-in type acoustical ceiling is considered __?__. (Review Article 100 Definitions of the *NEC*.)
    a. accessible
    b. exposed
    c. concealed
    d. both a and b

Codeology Fundamentals

This chapter introduces the fundamentals of the *Codeology* method. Previous chapters have introduced the basics of the four building blocks upon which the foundation of the *Codeology* method is based:

- Contents pages
- Section 90.3 *Code* arrangement
- Structure of the *NEC*
- Definitions, the language of the *NEC*

The *Codeology* method is described in detail by applying an outline of the *Codeology* method to the Contents page and describing the fundamental steps of this system. This chapter also describes how to customize your *Code* book; enhancing the *Codeology* method through the use of notes, highlighting, and tabs.

Objectives

» Identify the importance of customizing your *Code* book with notes, highlighting articles and parts, underlining, and applying *Code* tabs.

» Identify clues and key words to locate the proper chapter, article, and part within the *NEC*.

» Name the four basic building blocks of *Codeology*.

» Recognize the *Codeology* outline of the *NEC*.

» Implement the fundamentals of *Codeology*.

Chapter 5

Table of Contents

MARKING UP YOUR *CODE* BOOK

Because new copies of the *NEC* are not marked or annotated in any way, *Code* book users should personalize them by adding their name to the spine or to the inside cover. The *NEC* is not printed in color, so marking up and annotating your *Code* book will enhance the *Codeology* method. The following suggestions can make your copy of the *NEC* easier to use and will enhance your ability to quickly and accurately find needed information using the *Codeology* method.

Annotating the Contents Pages

The *NEC* Contents Pages are the starting point for researching all *Code* inquiries.

While each major subdivision of the Contents pages is titled, clarifying the scope with notes can be extremely helpful. **See Figure 5-1.** Begin the *Codeology* method by writing the following notes in the Contents pages to define the chapter scope:

Highlighting Your *Code* Book

Highlighters can be extremely useful tools, especially for *Code* users. The number one rule when highlighting is to highlight only where absolutely necessary. Some users of the *NEC* use highlighters liberally, sometimes highlighting an entire page. This counterproductive practice would force a user to read an entire page instead of a smaller, pertinent portion.

	The	
Article 90	Introduction *and ground rules for the NEC®*	
Chapter 1	General *Information and Rules for Electrical Installations*	
	Information and Rules on	
Chapter 2 *Plan*	Wiring and Protection *of Electrical Installations*	
	Information and Rules on	
Chapter 3 *Build*	Wiring Methods and Materials *for use in Electrical Installations*	
	Information and Rules on	
Chapter 4 *Use*	Equipment for General Use *for use in Electrical Installations*	
	Modifications and/or Supplemental Information/Rules for Electrical Installations in	
Chapter 5	Special Occupancies	
	Modifications and/or Supplemental Information/Rules for Electrical Installations containing	
Chapter 6	Special Equipment	
	Modifications and/or Supplemental Information/Rules for Electrical Installations containing	
Chapter 7	Special Conditions	
Chapter 8	Communications Systems *only*	

Simple notes can aid the user in quickly identifying the correct chapter to look for a Code reference.

Figure 5-1   Annotating the Content Pages

Figure 5-1. Marking up the NEC contents pages will help in the Codeology process.

Highlighters are available in a variety of colors. *Code* users should use those colors to their advantage. The most productive use of highlighters is to identify each article and part in order to have them stand out when moving through the *NEC*. The following exercise is mandatory for all *Codeology* students and requires green and orange highlighters. Green is used to highlight all articles and orange is used to highlight all parts. These colors are not to be used for any other highlighting in the *Code* book, which allows the *Codeology* user to instantly recognize the beginning of a new part or a new article. Color coding is extremely important, because it allows *Code* users to start in the correct part of the correct article and move quickly to the next part of the article to find an answer. While green and orange are used for marking only articles and parts, yellow should be used for highlighting other *Code* text, but only when necessary.

The following exercise requires a green and an orange highlighter:

1. Open your *NEC* to the Table of Contents.
2. Begin with Article 90 and determine the page number from the Table of Contents. Locate the page and highlight the article number and title in green.

ARTICLE 90 Introduction

3. Move on to Article 100 and determine the page number from the Table of Contents. Locate the page and highlight the article number and title in green, and the parts in orange, as follows:

ARTICLE 100 Definitions

I. General

II. Over 600 Volts, Nominal

4. Move on to Article 110 and determine the page number from the Table of Contents. Locate the page and highlight the article number and title in green, and the parts in orange, as follows:

ARTICLE 110 Requirements for Electrical Installations

I. General

II. 600 Volts, Nominal, or Less

III. Over 600 Volts, Nominal

IV. Tunnel Installations over 600 Volts, Nominal

V. Manholes and Other Electric Enclosures Intended for Personnel Entry, All Voltages

Continue through the Contents Pages, finding the page number for each article, locate the article and highlight it in green and locate each part of an article and highlight it in orange.

Marking each article in green and each part in orange will enable you to quickly identify all the articles and parts of articles. When applying the *Codeology* method, it is imperative to know at all times which part of which article a section is located in to properly apply the requirement. **See Figure 5-2.**

Applying *Code* Tabs

Code tabs are an extremely effective tool for quickly finding frequently used articles

---

Fact

The *NEC* is available as a PDF file for a low cost, or free as a promotion. Utilizing the search feature of a PDF reader, the user can search the entire *NEC* quickly. Try searching for "substation."

Figure 5-2   Highlighting the *Code* Book

408.58 Panelboard Marking. Panelboards shall he durably marked by the manufacturer with the voltage and the current rating and the number of phases for which they are designed and with the manufacturer's name or trademark in such a manner so as to be visible after installation, without disturbing the interior parts or wiring.

ARTICLE 409
Industrial Control Panels

I. General

48V, I Scope, This article covers industrial control panels intended for general use and operating at 600 volts, or less.

> Informational Note: UL 508A-2001. *Standard for Industrial Control Panels,* is a safety standard for industrial control panels.

409.2 Definitions.

Control Circuit. The circuit of a control apparatus or system that carries the electric signals directing the performance of the controller but does not carry the main power current.

Industrial Control Panel. An assembly of two or more components consisting of one of the following:

(1) Power circuit components only, such as motor controllers, overload relays, fused disconnect switches, and circuit breakers

(2) Control circuit components only, such as pushbuttons, pilot lights, selector switches, timers, switches, control relays

(3) A combination of power and control circuit components

These components, with associated wiring and terminals, are mounted on or contained within an enclosure or mounted on a subpanel. The industrial control panel does not include the controlled equipment.

409.3 Other Articles, In addition to the requirements of Article 409, industrial control panels that contain branch circuits for specific loads or components, or are for control of specific types of equipment addressed in other articles of this *Code,* shall be constructed and installed in accordance with the applicable requirements from the specific articles in Table 409.3.

II. Installation

409.20 Conductor — Minimum Size and Ampacity. The size of the industrial control panel supply conductor shall

Table 409.3 Other Articles

Equipment/Occupancy	Article	Section
Branch circuits	210	
Luminaires	410	
Motors, motor circuits, and controllers	430	
Air-conditioning and refrigerating equipment	440	
Capacitors		460.8, 460.9
Hazardous (classified) locations	500, 501, 502, 503, 504, 505	
Commercial garages; aircraft hangars; motor fuel dispensing facilities: bulk storage plants.: spray application, dipping, and outing processes; and inhalation anesthetizing locations	511, 513, 514, 515, 516, and 517 Part IV	
Cranes and hoists	610	
Electrically driven or controlled irrigation machines	675	
Elevators, dumbwaiters. escalators, moving walks. wheelchair lifts, and stairway chair lifts	620	
Industrial machinery	670	
Resistors and reactors	470	
Transformers	450	
Class 1. Class 2, and Class 3 remote-control, signaling, and power-limited circuits	725	

have un ampacity not less than 125 percent of the full-load current rating of all resistance heating loads plus 125 percent of the kill-load current rating of the highest rated motor plus the sum of the full-load current ratings of all other connected motors and apparatus based on their duty cycle that may be in operation at the same time,

409.21 Overcurrent Protection.

(A) General. Industrial control panels shall be provided with overcurrent protection in accordance with Parts I, II. and IX of Article 240.

(B) Location. This protection shall be provided for each incoming supply circuit by either of the following:

(1) An overcurrent protective device located ahead of the industrial control panel.

(2) A single main overcurrent protective device located within the industrial control panel. Where overcurrent protection is provided as part of the industrial control panel, the supply conductors shall be considered as either feeders, or taps as covered by 240.21.

Figure 5-2. The Codeology method includes highlighting the article titles with green and the part titles with orange.

Figure 5-3 | Underlining Key Material

(D) Secured. Service-entrance cables shall be held securely in place.

(E) Separately Bushed Openings. Service heads shall have conductors of different potential brought out through separately bushed openings.

Exception: For jacketed multiconductor service-entrance cable without splice.

(F) Drip Loops. Drip loops shall be formed on individual conductors. To prevent the entrance of moisture, service-entrance conductors shall be connected to the service-drop or overhead service conductors either (1) below the level of the service head or (2) below the level of the termination of the service-entrance cable sheath.

(G) Arranged That Water Will Not Enter Service Raceway or Equipment. Service-entrance and overhead service conductors shall be arranged so that water will not enter service raceway or equipment.

230.56 Service Conductor with the Higher Voltage to Ground. On a 4-wire, delta-connected service where the midpoint of one phase winding is grounded, the service conductor having the higher phase voltage to ground shall be durably and permanently marked by an outer finish that is orange in color, or by other effective means, at each termination or junction point.

V. Service Equipment — General

230.62 Service Equipment — Enclosed or Guarded. Energized ports of service equipment shall be enclosed as specified in 230.62(A) or guarded as specified in 230.62(B).

(A) Enclosed. Energized parts shall be enclosed so that they will not be exposed to accidental contact or shall be guarded as in 230.62(B).

(B) Guarded. Energized parts that are not enclosed shall be installed on a switchboard, panelboard, or control board and guarded in accordance with 110.18 and 110.27. Where energized parts are guarded as provided in 110.27(A)(1) and (A)(2), a means for locking or sealing doors providing access to energized parts shall be provided.

230.66 Marking. Service equipment rated at 600 volts or less shall be marked to identify it as being suitable for use as service equipment. All service equipment shall be listed. Individual meter socket enclosures shall not be considered service equipment.

VI. Service Equipment — Disconnecting Means

230.70 General. Means shall be provided to disconnect all conductors in a building or other structure from the service-entrance conductors.

(A) Location. The service disconnecting means shall be installed in accordance with 230.70(A)(1), (A)(2), and (A)(3).

(1) Readily Accessible Location. The service disconnecting means shall be installed at a readily accessible location either outside of a building or structure or inside nearest the point of entrance of the service conductors.

(2) Bathrooms. Service disconnecting means shall not be installed in bathrooms.

(3) Remote Control. Where a remote control device(s) is used to actuate the service disconnecting means, the service disconnecting means shall be located in accordance with 230.70(A)(1).

(B) Marking. Each service disconnect shall be permanently marked to identify it as a service disconnect.

(C) Suitable for Use. Each service disconnecting means shall be suitable for the prevailing conditions. Service equipment installed in hazardous (classified) locations shall comply with the requirements of Articles 500 through 517.

230.71 Maximum Number of Disconnects.

(A) General. The service disconnecting means for each service permitted by 230.2, or for each set of service-entrance conductors permitted by 230.40, Exception No. 1, 3, 4, or 5, shall consist of not more than six switches or sets of circuit breakers, or a combination of not more than six switches and sets of circuit breakers, mounted in a single enclosure, in a group of separate enclosures, or in or on a switchboard. There shall be not more than six sets of disconnects per service grouped in any one location.

For the purpose of this section, disconnecting means installed as part of listed equipment and used solely for the following shall not be considered a service disconnecting means:

(1) Power monitoring equipment
(2) Surge-protective device(s)
(3) Control circuit of the ground-fault protection system
(4) Power-operable service disconnecting means

(K) Single-Pole Units. Two or three single-pole switches or breakers, capable of individual operation, shall be permitted on multiwire circuits, one pole for each ungrounded conductor, as one multipole disconnect, provided they are equipped with identified handle ties or a master handle to disconnect all conductors of the service with no more than six operations of the hand.

Informational Note: See 408.36, Exception No. 1 and Exception No. 3, for service equipment in certain panelboards, and see 430.95 for service equipment in motor control centers.

Figure 5-3. Along with highlighting, underlining will aid in the Codeology process.

and sections of the *NEC*. When applying *Code* tabs, be sure to take the time to apply them properly. Read and follow the instructions provided with your tabs.

Underlining Key Material

Another effective practice when using the *NEC* is to underline sections, subdivisions, list items, or any material you want to bring to your attention in the future. Together with highlighting, underlining important text is extremely useful when revisiting a specific area of the *NEC*. **See Figure 5-3.**

Notes for Specific Section Application

As your *Code* studies progress to article-specific or topic-specific courses, take the time to write brief, neat notes in pencil for future reference. **See Figure 5-4.** Use the blank pages in the back of the *NEC* for additional notes. Those blank pages are ideal for making notes for possible proposals to change the next edition of the *NEC*. If you are planning on taking a qualifying exam for a State Electrical License, know that some States do not allow written notes in an NEC book used for the exam. Check with your State and see if this rule applies for your area. If the test proctor finds notes in your book you may have to obtain a "clean" version of the NEC for your test.

CLUES AND KEY WORDS

Clues come in many forms and each step of the *Codeology* method exposes clues for individual chapters, articles, and parts. The use of clues or key words is essential when qualifying a question or when you need to get to the correct part of the correct article in the correct chapter. Clues are found throughout the following:
• Questions
• Chapter titles
• Article titles
• Titles of parts
• Titles of sections and subdivisions

Clues or key words include, but are not limited to, the following:

Basic/General	Chapter 1
Plan	Chapter 2
Build	Chapter 3
Use	Chapter 4
Specials	Chapters 5 - 7
Communications	Chapter 8
Occupancy Types	Chapters 1 - 7

Indoor	Outdoor
Dry Location	Damp/Wet
Permanent	Temporary
600 Volts or Less	Over 600 Volts
Ungrounded	Grounded
Shall	Shall Not

Clues and key words will be examined in subsequent chapters of the *Codeology* text, with focus being placed on each chapter in the Contents Pages.

FOUR BASIC BUILDING BLOCKS OF THE *CODEOLOGY* METHOD

The previous chapters established the foundation for introducing and applying the *Codeology* method. Four basic building blocks form the basis of our *Codeology* method and can be summarized as outlined below:

Building Block #1
Contents Pages
• Outline of the *NEC*
• Ten major subdivisions of the *NEC*
  • Introduction, **Article 90**
  • Chapters 1 through 9
• Subdivision of chapters into articles within the scope of each chapter

Figure 5-4	Notes for Specific Section Application

Controller. For the purpose of this article, a controller is any switch or device that is normally used to start and stop a motor by making and breaking the motor circuit current.

Motor Control Circuit. The circuit of a control apparatus or system that carries the electric signals directing the performance of the controller but does not carry the main power current.

System Isolation Equipment. A redundantly monitored, remotely operated contactor-isolating system, packaged to provide the disconnection/isolation function, capable of verifiable operation from multiple remote locations by means of lockout switches, each having the capability of being padlocked in the "off" (open) position.

Valve Actuator Motor (VAM) Assemblies. A manufactured assembly, used to operate a valve, consisting of an actuator motor and other components such as controllers, torque switches, limit switches, and overload protection.

> Informational Note: VAMs typically have short-time duty and high-torque characteristics.

430.4 Part-Winding Motors. A part-winding start induction or synchronous motor is one that is arranged for starting by first energizing part of its primary (armature) winding and, subsequently, energizing the remainder of this winding in one or more steps. A standard part-winding start induction motor is arranged so that one-half of its primary winding can be energized initially, and, subsequently, the remaining half can be energized, both halves then carrying equal current. A hermetic refrigerant compressor motor shall not be considered a standard part-winding start induction motor.

Where separate overload devices are used with a standard part-winding start induction motor, each half of the motor winding shall be individually protected in accordance with 430.32 and 430.37 with a trip current one-half that specified.

Each motor-winding connection shall have branch-circuit short-circuit and ground-fault protection rated at not more than one-half that specified by 430.52.

Exception: A short-circuit and ground-fault protective device shall be permitted for both windings if the device will allow the motor to start. Where time-delay (dual-element) fuses are used, they shall be permitted to have a rating not exceeding 150 percent of the motor full-load current.

430.5 Other Articles. Motors and controllers shall also comply with the applicable provisions of Table 430.5.
* Always start at 430.6

430.6 Ampacity and Motor Rating Determination. The size of conductors supplying equipment covered by Article 430 shall be selected from the allowable ampacity tables in accordance with 310.15(B) or shall be calculated in accordance with 310. 15(C). Where flexible cord is used, the size

Table 430.5 Other Articles

Equipment/Occupancy	Article	Section
Air-conditioning and refrigerating equipment	440	
Capacitors		460.8, 460.9
Commercial garages; aircraft hangars; motor fuel dispensing facilities; bulk storage plants; spray application, dipping, and coating processes; and inhalation anesthetizing locations	511, 513, 514, 515, 516, and 517 Part IV	
Cranes and hoists	610	
Electrically driven or controlled irrigation machines	675	
Elevators, dumbwaiters, escalators, moving walks, wheelchair lifts, and stairway chair lifts	620	
Fire pumps	695	
Hazardous (classified) locations	500–503 and 505	
Industrial machinery	670	
Motion picture projectors		540.11 and 540.20
Motion picture and television studios and similar locations	530	
Resistors and reactors	470	
Theaters, audience areas of motion picture and television studios, and similar locations		520.48
Transformers and transformer vaults	450	

of the conductor shall be selected in accordance with 400.5-. The required ampacity and motor ratings shall be determined as specified in 430.6(A), (B), (C), and (D).

(A) General Motor Applications. For general motor applications, current ratings shall be determined based on (A)(1) and (A)(2).

(1) Table Values. Other than for motors built for low speeds (less than 1200 RPM) or high torques, and for multispeed motors, the values given in Table 430.247, Table 430.248, Table 430.249, and Table 430.250 shall be used to determine the ampacity of conductors or ampere ratings of switches, branch-circuit short-circuit and ground-fault protection, instead of the actual current rating marked on the motor nameplate. Where a motor is marked in amperes, but not horsepower, the horsepower rating shall be assumed to be that corresponding to the value given in Table 430.247, Table 430.248, Table 430.249, and Table 430.250, interpo-

Figure 5-4. Writing notes in the Code book will ensure clarity for future reference.

Building Block #2
Section 90.3 Code Arrangement

- Chapters 1 through 4 apply generally to all installations
- Chapters 5, 6, and 7 supplement or modify Chapters 1 through 4
- Chapter 8 stands alone unless a specific reference exists
- Chapter 9 tables apply as referenced; Annexes are informational only

Building Block #3
Structure of the *NEC*

- Ten major subdivisions of the *NEC*
  - Introduction, **Article 90**
  - Chapters 1 through 9
- Broad area of coverage in each chapter
- Subdivision of each chapter into articles to address chapter scope
- Logical subdivision of articles into parts
- Subdivision of parts into sections
- Sections can contain three levels of subdivision
- Sections and subdivisions can contain exceptions
- Sections and subdivisions can contain list items
- Sections and subdivisions can contain informational notes
- Mandatory language: *shall* or *shall not*
- Permissive language: *shall be permitted* or *shall not be required*
- Informational material contained in informational notes and Annexes

Building Block #4
Definitions, the Language of the *NEC*

- Terms used in more than one article are defined in **Article 100**
- Terms used in a single article are defined in the second section of the article

GETTING TO THE CORRECT CHAPTER, ARTICLE, AND PART

By design, the *Codeology* method allows users to determine exactly where to begin looking in the *NEC* for the section which addresses their needs. This method begins by applying the *Codeology* outline to get to the correct chapter. Through the use of key words and clues, users can then identify the correct article and part in which to begin their inquiry into the *NEC*. These are the fundamentals of *Codeology*.

The *Codeology* method is designed to teach a systematic, disciplined approach to quickly find information by understanding and applying the outline form of the *NEC*. The use of generic terms to aid the *Codeology* user in finding the correct chapter in the contents pages is essential. The generic terms include Plan, Build, and Use. This portion of the chapter will briefly introduce the basic steps of *Codeology*, which will then be explained in detail in later chapters.

The following subsections explain the basic steps to finding needed information in the *NEC* quickly and accurately. This outline uses several general terms to help qualify the question or need in order to quickly identify the correct chapter. Clues and key words also help identify that chapter.

The general terms used to guide the *Codeology* user in the correct direction are Basic/General, Plan, Build, Use, Specials, and Communications. The primary focus is on *NEC* Chapters 1 through 8. **Article 90 Introduction** lays the ground rules for the use of the *NEC*, while the tables in Chapter 9 are to be used only where referenced elsewhere in the *Code* book. The use of these terms in the *Codeology* method is described next.

Basic/General

When a question or need for information in the *NEC* is basic or general in nature to all electrical installations, think "General"

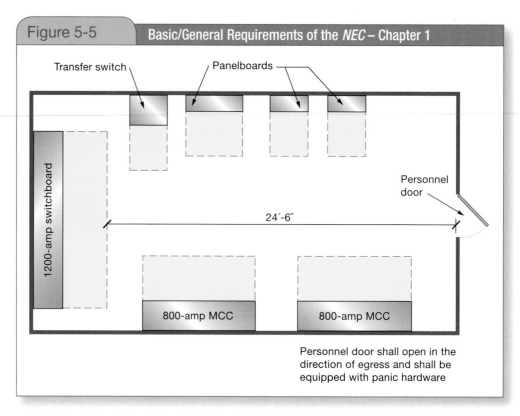

Figure 5-5. The general requirements of Chapter 1 include provisions for adequate working space for persons and dedicated space for electrical equipment.

and start in **Chapter 1 General**. Chapter 1 contains two articles: Definitions and Requirements for Electrical Installations.

See Figure 5-5. Key words and clues which should guide the *Codeology* user to think "General, Chapter 1" include the following:

- A definition question
- Examination, installation, and use of equipment (listed, labeled)
- Mounting and cooling of equipment
- Electrical connections
- Flash protection
- Identification of disconnecting means
- Workspace clearance

Plan

When a question or need for information in the *NEC* deals with planning stages and are general in nature to all electrical installations, think "Plan" and start in **Chapter 2 Wiring and Protection**. **See Figure 5-6.** The title of Chapter 2 includes two key words: wiring and protection.

The term "wiring," as used in Chapter 2, does not imply different types of cable assemblies or raceways. Wiring in

Figure 5-6. Chapter 2 of the NEC is called the "Plan" chapter. All electrical installations must be properly planned.

Chapter 2 addresses the *NEC* terms for the types of current-carrying conductors common to all electrical installations. These wiring types include branch circuits, feeders, services, taps, and transformer secondary conductors. Chapter 2 provides the basic requirements for all of these conductors and the calculations to properly size them.

The term "protection," as used in Chapter 2, includes the basic protection requirements for all electrical installations, including overcurrent protection, grounding, bonding, surge arresters, and surge-protective devices.

Key words and clues which should guide the *Codeology* user to think "Plan, Chapter 2" include the following:

- Use and identification of grounded conductors
- Branch circuit
- Feeder
- Service
- Calculation or computed load
- Overcurrent protection
- Grounding
- Surge arresters
- Surge-protective devices (SPDs), 1kV or Less

Build

When a question or need for information in the *NEC* deals with building an electrical installation, think "Build" and start in Chapter 3. All inquiries regarding methods and materials to get from the source of energy to the load (all physical wiring of an installation) require wiring methods and/or materials covered in **Chapter 3 Wiring Methods and Materials**. The title of Chapter 3 includes two terms: wiring methods and wiring materials, which cover all means and methods of electrical distribution. Chapter 3 does not cover panelboards, disconnects, transformers, or utilization equipment. But, it does cover every physical means of wiring branch circuits, feeders, services, taps, and transformer secondary conductors.

The wiring methods portion of Chapter 3 includes general information for all wiring methods and materials as well as specific articles for all conductors, cable assemblies, raceways, busways, cablebus, and the like. In addition, the wiring materials portion of Chapter 3 includes general information for all wiring materials as well as specific articles for all boxes (outlet, device, pull, junction), conduit

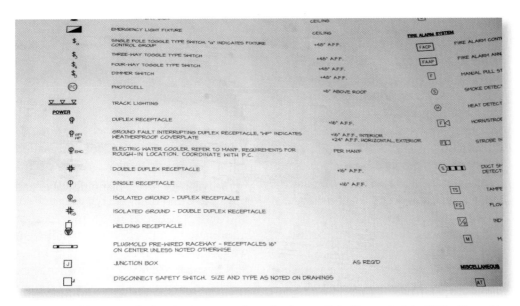

Symbols on a blueprint will identify what materials need to be installed. Chapter 3 of the NEC will provide the Code requirements for the installation.

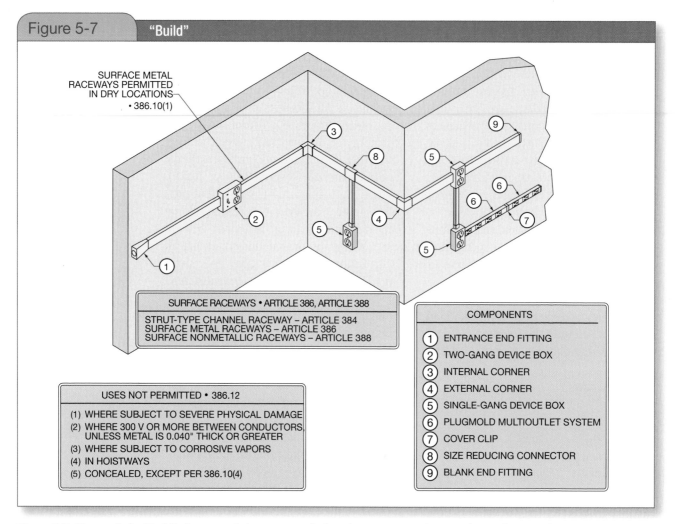

Figure 5-7   "Build"

SURFACE METAL
RACEWAYS PERMITTED
IN DRY LOCATIONS
• 386.10(1)

SURFACE RACEWAYS • ARTICLE 386, ARTICLE 388

STRUT-TYPE CHANNEL RACEWAY – ARTICLE 384
SURFACE METAL RACEWAYS – ARTICLE 386
SURFACE NONMETALLIC RACEWAYS – ARTICLE 388

USES NOT PERMITTED • 386.12

(1) WHERE SUBJECT TO SEVERE PHYSICAL DAMAGE
(2) WHERE 300 V OR MORE BETWEEN CONDUCTORS,
     UNLESS METAL IS 0.040" THICK OR GREATER
(3) WHERE SUBJECT TO CORROSIVE VAPORS
(4) IN HOISTWAYS
(5) CONCEALED, EXCEPT PER 386.10(4)

COMPONENTS

(1) ENTRANCE END FITTING
(2) TWO-GANG DEVICE BOX
(3) INTERNAL CORNER
(4) EXTERNAL CORNER
(5) SINGLE-GANG DEVICE BOX
(6) PLUGMOLD MULTIOUTLET SYSTEM
(7) COVER CLIP
(8) SIZE REDUCING CONNECTOR
(9) BLANK END FITTING

Figure 5-7. Chapter 3, the "Build" chapter, includes wiring methods and wiring materials to get electrical current from point "A" to point "B" in all electrical installations.

For additional
information, visit
qr.njatcdb.org
Item #1064

bodies, cabinets, cutout boxes, auxiliary gutters, and wireways. **See Figure 5-7.** Key terms and clues which should guide the *Codeology* user to think "Build, Chapter 3" include the following:

• General questions on wiring methods or materials
• General questions for conductors, uses permitted, ampacity, etc.
• Installation, uses permitted, or construction of any cable assembly
• Installation, uses permitted, or construction of any raceway
• Installation, uses permitted, or construction of any other distribution method
• Requirements for boxes of all types

• Requirements for cabinets, meter sockets, wireways, etc.
• Open wiring on insulators and outdoor overhead conductors over 600 volts

Use

When a question or need for information in the *NEC* deals in general with electrical equipment which uses, controls, or transforms electrical energy, think "Use" and start in Chapter 4. All inquiries for information on electrical equipment which controls, transforms, utilizes, or aids in the utilization of electrical energy can be found in **Chapter 4 Equipment for General Use**.

Chapter 4 covers equipment which uses electrical energy to perform a task such as luminaires (lighting fixtures), appliances, air conditioners, motors, and heating, de-icing, and snow-melting equipment. In addition, Chapter 4 deals with equipment which controls or facilitates the use of electrical energy, such as switches, switchboards, panelboards, cords, cord caps, and receptacles. Chapter 4 also covers electrical equipment which generates or transforms electrical energy, such as generators, transformers, batteries, phase converters, capacitors, resistors, and reactors. **See Figure 5-8.** Among the key words and clues which should guide the *Codeology* user to think "Use, Chapter 4" are the following:

- Cords or cables
- Luminaires (lighting fixtures and fixture wires)
- Receptacles and switches
- Switchboards and panelboards
- Appliances
- Heating equipment
- Air-conditioning equipment
- Motors
- Batteries
- Generators
- Transformers
- Phase converters
- Capacitors, resistors, and reactors
- Storage batteries

Figure 5-8. Chapter 4 of the NEC is called the "Use" chapter. Utilization equipment, including motors, is covered in this chapter along with articles dedicated to the control and installation of utilization equipment.

The Specials

When a question or need arises for information dealing with special occupancies, equipment, or conditions, think "Special" and start in Chapters 5, 6, or 7. **See Figure 5–9.** Chapters 1 through 4 are the foundation or backbone for all electrical installations. For installations which contain a "Special Occupancy, Equipment, or Condition," Chapters 5, 6, or 7 supplement or modify the first four chapters to address the "Special" situation. Key words and clues which should steer the *Codeology* user to think "Special Occupancies, Chapter 5" include the following:

- Hazardous, classified locations

Figure 5-9. Chapters 5, 6, and 7 of the NEC are the "Special" chapters. Chapter 5 covers Special Occupancies (including hazardous locations), Chapter 6 covers Special Equipment (including spas), and Chapter 7 covers Special Conditions (including emergency systems).

- Commercial garages, motor fuel dispensers
- Spray booths and applications
- Hospitals and all health care facilities
- Places of assembly
- Theaters
- Carnivals
- Temporary power
- Manufactured buildings, motor homes, RVs
- Agricultural buildings, farms
- Marinas, floating buildings

Key words and clues which should guide the *Codeology* user to think "Special Equipment, Chapter 6" include the following:

- Electric signs
- Cranes
- Elevators
- Electric welders
- X-ray equipment
- Swimming pools
- Solar and fuel cell systems
- Fire pumps

Key words and clues which should guide the *Codeology* user to think "Special Conditions, Chapter 7" include the following:

- Emergency systems
- Legally required and optional standby systems
- Class 1, 2, and 3 systems
- Fire alarm systems
- Fiber optics

Communications

When a question or need for information in the *NEC* deals with communications systems, think *Communication* and start in Chapter 8. Chapter 8 stands alone, in that the rest of the *NEC* does not apply to this chapter unless specifically referenced as such within Chapter 8. **See Figure 5-10.** Key words and clues which should guide the *Codeology* user to think "Communications, Chapter 8" include the following:

Communications circuits

- Radio and TV
- CATV systems

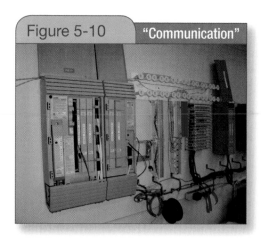

Figure 5-10. *Chapter 8 of the NEC covers communications systems.*

- Network powered broadband systems

The *Codeology* outline provides a simplified view of all *NEC* chapters and how they are placed in specific groups when using the *Codeology* method. **See Figure 5-11.**

FUNDAMENTAL STEPS TO USING *CODEOLOGY*

When a need arises to find information in the *NEC*, several steps may be necessary to find the answers. The following are situations or circumstances for which such information may be sought:

- Job site problems/questions
- Design concerns
- Inspection questions
- *NEC* proficiency exam

The following steps should be followed to get your problems/questions answered:

Step 1

- Qualify your question or need
- Go to the Contents pages
- Look for clues or key words to go to the correct chapter
- Get to the correct chapter

Step 2

- Further qualify your question or need
- Look for clues or key words
- Use the Contents pages to get to the correct article within the correct chapter

Figure 5-11 | The *Codeology* Method

THE GROUND RULES		
Introduction	Article 90	Introduction/Directions
THE BASIC INSTALLATION		
Chapter 1	100 Series	General
Chapter 2	200 Series	Plan
Chapter 3	300 Series	Build
Chapter 4	400 Series	Use
THE SPECIALS		
Chapter 5	500 Series	Occupancies
Chapter 6	600 Series	Equipment
Chapter 7	700 Series	Conditions
COMMUNICATIONS: THE LONER		
Chapter 8	800 Series	Communications
TABLES		
Chapter 9	Tables and Annexes	

Figure 5-11. The Codeology method is broken down into five groups to aid the Code user in finding information in the NEC.

Step 3
- Further qualify your question or need
- Look for clues or key words
- Use the Contents pages to go into the correct part of the correct article

Step 4
- Open the *NEC* to the part of the article which meets your question or need

Step 5
- Read only the section titles to find the correct section

Step 6
- Read section title and all titles of first-level subdivisions
- Read all of the section and pertinent subdivisions including exceptions and Informational Notes
- Apply the rule or answer your question

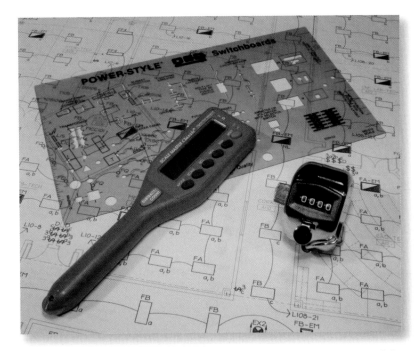

A wire pull requires the Electrical Worker to know how much wire will be needed. The installation requires the Electrical Worker to understand the Code requirements.

Summary

Learning, understanding, and applying the *Codeology* method begins with four basic building blocks:

- *Building block #1*, the **Contents** pages: The *Code* user must be familiar with and understand the 10 major subdivisions of the *NEC*.
- *Building block #2*, the **Arrangement** of the *NEC* as detailed in **Section 90.3**: The *Code* user must understand how the individual chapters of the *NEC* apply.
- *Building block #3*, the structure or **Outline** form of the *NEC*: Without an understanding of the hierarchy of requirements and information in the *NEC*, proper application would be impossible.
- *Building block #4*, the **Language** of the *NEC*: An in-depth understanding of defined terms is essential for proper application of *NEC* requirements.

The *Codeology* outline builds upon the arrangement of the *NEC* and provides a user-friendly method to help categorize or qualify a question or need through key words and clues. The *Codeology* outline is as follows:

- **The Ground Rules**
  - Introduction   Article 90   Introduction/Directions
- **The Basic Installation**
  - Chapter 1   100-series   General
  - Chapter 2   200-series   Plan
  - Chapter 3   300-series   Build
  - Chapter 4   400-series   Use
- **The Specials**
  - Chapter 5   500-series   Occupancies
  - Chapter 6   600-series   Equipment
  - Chapter 7   700-series   Conditions
- **Communications, Stand-Alone**
  - Chapter 8   800-series   Communications
- **Tables**
  - Chapter 9   Tables and Annexes

Understanding the *Codeology* outline along with a solid foundation in the four basic building blocks allows the *Code* user to begin to apply the *Codeology* method. Additional fundamentals include recognizing key words and clues and marking up the *Code* book by highlighting, underlining, inserting tabs, and making notes where appropriate. All of these fundamentals must be solidly in place before the *Codeology* method is applied to the introduction and nine chapters of the *NEC*.

1. *National Electrical Code* users can customize their *Code* book by which of the following methods?
   a. Highlighting
   b. Making notes and underlining
   c. Using *Code* tabs
   d. All of the above

2. Using the *Codeology* method, *Code* users must determine the proper clues or key words and go to the __?__.
   a. most likely article
   b. Contents pages
   c. index
   d. highlighted areas

3. After using key words or clues to get into the correct chapter of the *NEC*, *Codeology* users must further qualify their needs and go to the correct __?__.
   a. section
   b. part
   c. article
   d. subdivision

4. After using key words or clues to get to the correct chapter and then the correct article of the *NEC*, *Codeology* users must further qualify their needs and go to the correct __?__.
   a. section of the part
   b. part of the article
   c. table
   d. subdivision

5. Which chapter of the *NEC* is considered the "stand-alone" one?
   a. 7
   b. 8
   c. 9
   d. 10

6. An article in the 600 series will deal with special __?__.
   a. permission
   b. occupancies
   c. equipment
   d. none of the above

7. The *Codeology* method will allow the *Code* user to become __?__ with the *NEC*.
   a. accurate
   b. fast
   c. confident
   d. all of the above

8. The *Codeology* method requires that all articles be highlighted in green and all parts be highlighted in __?__ for fast identification when using the *NEC*.
   a. yellow
   b. pink
   c. blue
   d. orange

Article 90, The Introduction to the *NEC*

The Introduction to the *NEC* (**Article 90**) must be read and understood first before proceeding to any of the nine chapters which follow, and especially before applying the language of the *Code*. It is paramount to review and fully understand **Article 90** before the purchase and installation of equipment or appliances. To properly use an appliance or any equipment to achieve the desired results, the consumer must first understand the intent of the *Code*, what it covers and does not cover, how it is enforced, and when the language within the *Code* is mandatory, permissive, or just explanatory.

Objectives

» Identify **Article 90** as the article which contains directions essential for the Code user to understand and properly apply the requirements of the *NEC*.

» Understand that the ground rules contained in **Article 90** govern the entire *NEC*.

» Recognize that the purpose, scope, and arrangement of the *NEC* are governed by **Article 90**.

Chapter 6

Table of Contents

NEC: HOW TO BEGIN

It would be helpful for all users of the *National Electrical Code* if a warning was posted at the beginning of **Article 90. See Figure 6-1.** For newcomers, it is tempting to not fully review **Article 90** and to dive directly into Chapters 1– 9, where all the interesting electrical installations are discussed. But, not reviewing and understanding **Article 90** would leave the installer with the lack of knowledge of the intent of the *Code*, what it covers, and when the rules apply.

Article 90 is broken down into nine sections. **See Figure 6-2.** These sections discuss introductory information about the *NEC* including the ground rules, what applies, and who enforces the *NEC*.

ARTICLE 90 INTRODUCTION

Nine sections are contained in **Article 90**, instructions to the *NEC*. While **Article 90** only spans a few pages in the *NEC* and may seem unimportant, it provides the *Code* user with an understanding of how the *Code* should be applied along with various foundational concepts of the *Code* language.

Section 90.1 Purpose: Introduction to the *NEC*

Since the conception of the NFPA in 1896, safety has always be its primary purpose; this especially stands true for the *NEC*. These safety-driven requirements are intended to protect both persons and property through the standardization of safe installation practices.

Figure 6-1	Warning to Newcomers to the *NEC*

WARNING

READ AND UNDERSTAND THESE INSTRUCTIONS (Article 90) BEFORE ATTEMPTING TO USE THIS CODE. FAILURE TO READ AND UNDERSTAND THESE INSTRUCTIONS WILL RESULT IN THE FOLLOWING:

- Lack of understanding of the purpose of the *NEC*

- Attempts to enforce the *NEC* where it does not apply

- Lack of enforcement where the *NEC* does apply

- Lack of understanding of how the *NEC* provisions relate to the factory-installed wiring of electrical equipment

- Misapplication due to a lack of understanding of the arrangement of the *NEC*

- Misapplication due to a lack of understanding of the use of mandatory/permissive rules and explanatory material

- Misapplication of the *NEC* as a design manual

- Misapplication of the *NEC* as an instruction manual

- Misapplication due to a lack of understanding that the rules in the *NEC* are minimum standards

Figure 6-1. The purpose of Article 90 is to explain how the NEC is applied to electrical installations in the practical hazard safeguarding of people and property.

Figure 6-2	Nine Sections of Article 90
NEC® Title:	**Introduction**
Codeology Title:	**The Ground Rules**
Article Scope:	**The Introduction and Ground Rules for the NEC®**
Section	**Section Title**
90.1	Purpose
90.2	Scope
90.3	Code Arrangement
90.4	Enforcement
90.5	Mandatory Rules, Permissive Rules and Explanatory Material
90.6	Formal Interpretations
90.7	Examination of Equipment for Safety
90.8	Wiring Planning
90.9	Units of Measurement

Figure 6-2. Article 90 sections explain the foundation of the NEC language and how it should be applied.

90.1(A) Practical Safeguarding - The purpose of the *Code* is the practical safeguarding of persons and property from hazards arising from the use of electricity. **See Figure 6-3**.

90.1(B) Adequacy - Article 90 explains that the *NEC* contains the minimum requirements for safe wiring or installation of equipment. This section also warns that the use of the *NEC* does not guarantee an efficient, convenient, expandable,

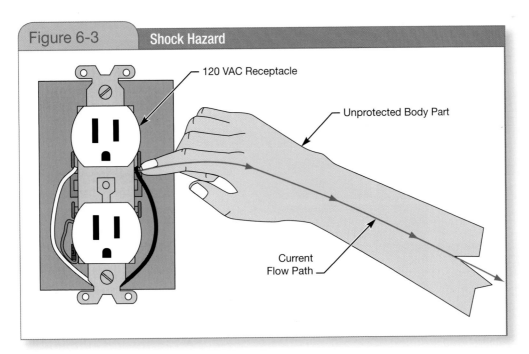

Figure 6-3	Shock Hazard

120 VAC Receptacle

Unprotected Body Part

Current Flow Path

Figure 6-3. Section 90.1 states that the purpose of the NEC is practical safeguarding of persons and equipment from hazards of electrical utilization.

or adequate installation. These are design issues which must be considered in addition to the *Code's* minimum installation requirements.

90.1(B) includes an Informational Note which generally states that often times hazards occur when wiring systems, which were at one time adequate, experience an increase in the usage of electricity higher than originally expected, resulting in the wiring becoming overloaded. An initial adequate installation with reasonable provisions for increased system usage should be realized for the future of the wiring. **See Figure 6-4.**

90.1(C) Intention - The *NEC* is not intended as a design specification or an instruction manual for untrained persons. Rather, it is an installation tool designed and revised every three years for use by trained persons in the electrical industry, inspectors, engineers, and manufacturers of electrical equipment.

90.1(D) Relation to Other International Standards - The *NEC* is also used as an international electrical installation standard. The text is intended to explain the relationship between the *NEC* and another electrical code used outside of the United States called Section 131 of *International Electrotechnical Commission (IEC) Standard 60364–1, Electrical Installations of Buildings.* In essence, the *NEC* addresses all of the protection requirements contained in the IEC Code.

Section 90.2 Scope: Application of the *NEC*

Section 90.2 sets the stage for applying the *NEC* in three subdivisions entitled "Covered," "Not Covered," and "Special Permission." It is important to review this section to determine when and where the *NEC* applies and when and where it does not. Note: Most electrical utilities use the NESC for installation guidelines for generation, transmission, distribution, and metering of services.

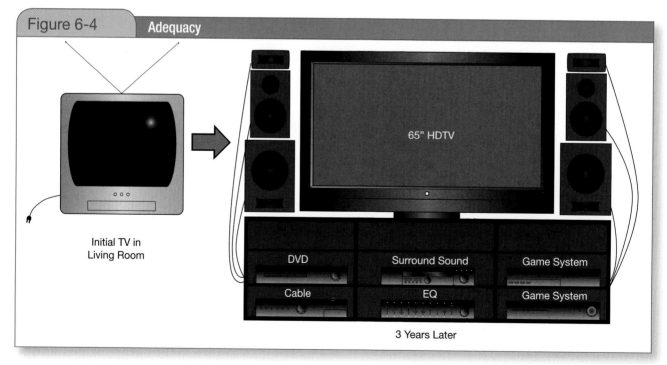

Figure 6-4. 90.1(B) Adequacy reminds the user that planning for future increased electricity usage should be part of the reasonable provisions of electrical installations.

90.2(A) Covered - The *NEC* covers the installation of the following:
- Electrical conductors
- Electric equipment
- Raceways
- Signaling and communications conductor equipment and raceways
- Optical fiber cables and raceways

The *NEC* covers the following premises and structures on the load side of the service point. **See Figure 6-5.**
- Public and private premises, including buildings, structures, mobile homes, recreational vehicles, and floating buildings
- Yards, lots, parking lots, carnivals, and industrial substations
- Installations of conductors and equipment that connect to the supply of electricity
- Installations used by the electric utility, such as office buildings, warehouses, garages, machine shops, and recreational buildings, that are not an integral part of a generating plant, substation, or control center

90.2(B) Not Covered - The *NEC* does not cover the following premises and structures:
- Installations in ships, watercraft other than floating buildings, railway rolling stock, aircraft, or automotive vehicles other than mobile homes and recreational vehicles
- Installations underground in mines and self-propelled mobile surface mining machinery and its attendant electrical trailing cable
- Installations of railways for generation, transformation, transmission, or distribution of power used exclusively for operation of rolling stock or installations used exclusively for signaling and communications purposes
- Installations of communications equipment under the exclusive control of communications utilities located out-

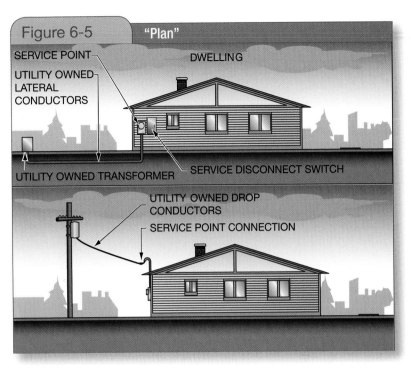

Figure 6-5   "Plan"

SERVICE POINT — DWELLING
UTILITY OWNED LATERAL CONDUCTORS
UTILITY OWNED TRANSFORMER — SERVICE DISCONNECT SWITCH
UTILITY OWNED DROP CONDUCTORS — SERVICE POINT CONNECTION

Figure 6-5. 90.2(A) Covered states that the NEC governs the conductors past the service point into the dwelling and not the utility owned lateral conductors from the transformer or the utility owned overhead conductors.

doors or in building spaces used exclusively for such installations
- Installations under the exclusive control of an electric utility where such installations:
  a. Consist of service drops or service laterals, and associated metering, or
  b. Are on property owned by or leased by the electric utility for the purpose of communications, metering, generation, control, transformation, transmission, or distribution of electric energy, or
  c. Are located in legally established easements or rights-of-way, or
  d. Are located by other written agreements either designated by or recognized by public service commissions, utility commissions, or other regulatory agencies having jurisdiction for such installations. These written agreements shall be limited to installations for the purpose of communications, metering, generations, control,

transformation, transmission, or distribution of electric energy where legally established easements or rights-of-way cannot be obtained. These installations shall be limited to federal lands, Native American reservations through the U.S. Department of the Interior Bureau of Indian Affairs, military bases, lands controlled by port authorities and State agencies and departments, and lands owned by railroads. **See Figure 6-6.**

90.2(C) Special Permission - The AHJ (authority having jurisdiction), as defined in **Article 100**, may grant special permission when service conductors used to connect to the utility are not under the control of the utility, provided the installation is located outside a building or structure, or terminate inside nearest the point of entrance of the service conductors.

Section 90.3 Code Arrangement: Application of Chapters

The *NEC* is arranged by chapter as follows:

General

Chapters 1 through 4 apply generally - meaning they apply at all times for every installation, in every occupancy unless supplemented or modified by Chapter 5, 6, or 7.

Special

Chapters 5, 6, and 7 address special requirements and supplement or modify Chapters 1 through 4.

Communications Systems

Chapter 8 stands alone. The remainder of the *NEC* does not apply to articles within Chapter 8 unless specifically referenced in Chapter 8. Broadcasting systems such as radio and television along with communicational systems are the primary focus.

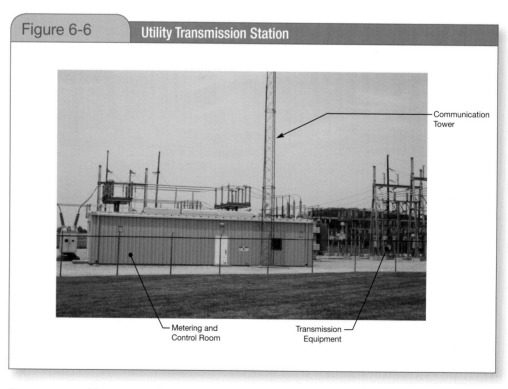

Figure 6-6 **Utility Transmission Station**

Communication Tower

Metering and Control Room

Transmission Equipment

Figure 6-6. 90.2(B) Not Covered states that facilities exclusively controlled by electrical utilities are not covered by the NEC.

Figure 6-7 — Code Arrangement

ARTICLE 90.3	
Chapters 1 through 4 apply GENERALLY to ALL electrical installations	
Chapter 1	GENERAL
Chapter 2	WIRING and PROTECTION
Chapter 3	WIRING METHODS and MATERIALS
Chapter 4	EQUIPMENT for GENERAL USE
Chapters 5, 6, and 7 SUPPLEMENT or MODIFY Chapters 1 through 4	
Chapter 5	SPECIAL OCCUPANCIES
Chapter 6	SPECIAL EQUIPMENT
Chapter 7	SPECIAL CONDITIONS
Chapter 8 is not subject to the requirements of Chapters 1 through 7 except where the requirements are specifically referenced in Chapter 8	
Chapter 8	COMMUNICATIONS SYSTEMS
Applicable as referenced	
Chapter 9	TABLES
Informational only; not mandatory	
Chapter 9	INFORMATIVE ANNEX A through INFORMATIVE ANNEX I

Figure 6-7. Section 90.3 states that the Code is divided into the introduction and nine chapters.

Tables

Tables located in Chapter 9 are applicable only when they are referenced in other chapters.

Annexes

Annexes A through I are for information-al purposes only. **See Figure 6-7.**

Section 90.4 Enforcement: of the *Code*

The *NEC* is intended to be adopted by, and enforced by, governmental bodies exercising legal jurisdiction over electrical installations, such as a city, state, township, county, or municipality. The authority having jurisdiction (AHJ), as defined in **Article 100**, is responsible for making interpretations and may give special permission for specific installations when it is determined that an equivalent level of installation safety can be achieved. **See Figure 6-8.**

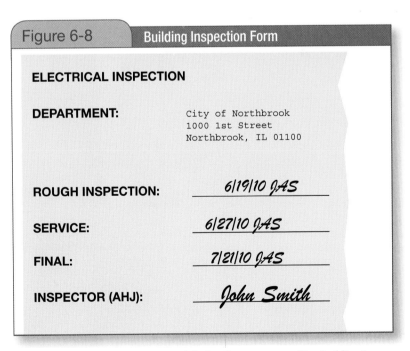

Figure 6-8 — Building Inspection Form

ELECTRICAL INSPECTION

DEPARTMENT: City of Northbrook
1000 1st Street
Northbrook, IL 01100

ROUGH INSPECTION: 6/19/10 JAS

SERVICE: 6/27/10 JAS

FINAL: 7/21/10 JAS

INSPECTOR (AHJ):   John Smith

Figure 6-8. Building Commissions of the local government utilize building inspectors as AHJ. Inspection forms are part of the documentation used throughout a project's construction period.

Section 90.5 Mandatory Rules, Permissive Rules, and Explanatory Material

The *NEC* contains mandatory and permissive rules along with explanatory material, which aids the *Code* user in understanding and applying the *NEC* requirements. **Section 90.5** provides clear instructions for determining which rules are mandatory or permissive and which are informational.

90.5(A) Mandatory Rules - A mandatory rule applies wherever the terms "shall" or "shall not" are used within the *NEC*.

90.5(B) Permissive Rules - A permissive rule applies wherever the terms "shall be permitted" or "shall not be required" are used within the *NEC*. These rules are not mandatory but are options or alternative methods.

90.5(C) Explanatory Materials - Explanatory material is included in the *NEC* in the form of Informational Notes, which are always written in small text below the section, subdivision, list item, or article to which they apply. Informational Notes are not an enforceable part of the *NEC*.

90.5(D) Informative Annexes - Informative Annexes contain nonmandatory information relative to the use of the *NEC*. Informative annexes are not part of the enforceable requirements of the *NEC*, but for informational purposes only. There are nine Informative Annexes, which include information related to Product Safety Standards, Ampacity Calculations, Conduit Fill, Types of Construction, and Tightening Torque Tables.

| Figure 6-9 | Control Panel Wiring |

Figure 6-9. Section 90.7 states that factory-installed internal wiring of listed equipment shall not be required to be inspected under the NEC.

Section 90.6 Formal Interpretations: Procedures for Processing

Formal interpretation procedures are established in the *NFPA Regulations Governing Committee Projects*. Initiation of a Formal Interpretation procedure requires a detailed written request for a formal interpretation through NFPA. This process is in place to promote uniform interpretation and application of the *NEC* in every city, state, township, county, or municipality in which it is adopted and enforced.

Section 90.7 Examination of Equipment for Safety

The *NEC* does not intend that factory-installed internal wiring of listed equipment be inspected when installed. When qualified electrical testing laboratories have determined that equipment meets the appropriate product standard and they list and label the equipment accordingly, an inspection is not required. **See Figure 6-9. Section 90.7** includes three Information Notes to better clarify suitability for installation in accordance with this *Code*.

> Informational Note No. 1: See requirements in 110.3.
>
> Informational Note No. 2: *Listed* is defined in Article 100.
>
> Informational Note No. 3: Informative Annex A contains an informative list of product safety standards for electrical equipment.

Section 90.8 Wiring Planning for Future Expansion of Wiring

The *NEC* does not require that an electrical installation be designed and constructed to allow for future expansion. It contains minimum safety-driven electrical installation requirements, not design elements for future expansion. **Section 90.8** provides informational text which explains that electrical installations, designed with future expansion as a consideration, allow for efficient, convenient, and safe expansion.

90.8(A) Future Expansion and Convenience - Design considerations which are essential for convenient future expansion of electrical installations include plans and specifications which require the following:

- Ample space in raceways
- Spare raceways
- Additional space for electric power and communications circuits
- Readily accessible distribution centers

90.8(B) Number of Circuits in Enclosures - 90.8(B) explains that throughout the *NEC*, the number of circuits permitted in enclosures of all types is restricted to varying degrees, depending upon the installation and type of enclosure. The purpose of this restriction is to minimize the damage which could occur to all circuits contained in an enclosure when one of those circuits experiences a ground fault or short circuit. **See Figure 6-10.** These restrictions also reduce the heating effect when multiple current-carrying conductors are grouped together.

Figure 6-10 | Spare Conduits

Figure 6-10. 90.8(A) Future Expansion and Convenience list design considerations for expansions such as extra conduit for future expansion.

Figure 6-11	Dual Units

Table 110.26 (A)(1) Working Spaces

Nominal Voltage to Ground	Minimum Clear Distance		
	Condition 1	Condition 2	Condition 3
0 – 150	914 mm (3 ft)	914 mm (3 ft)	914 mm (3 ft)
151 – 600	914 mm (3 ft)	1.07 m (3 ft 6 in.)	1.22 in. (4 ft)

Reprinted with permission from NFPA 70-2011, *National Electrical Code®*, Copyright© 2010, National Fire Protection Association, Quincy, MA 02169. This reprinted material is not the complete and official position of the NFPA on the referenced subject, which is represented only by the standard in its entirety.

Figure 6-11. Section 90.9 Units of Measurements specifies that the NEC uses both the metric and inch-pound system throughout the Code.

Section 90.9 Units of Measurement

When the *NEC* specifies distance, size, or weight, both metric and inch-pound units are used because the *NEC* is an international electrical installation standard. The United States still uses the inch-pound system while the rest of the world uses the metric system. **See Figure 6-11.**

90.9(A) Measurement System of Preference - The metric units used in the *NEC* are in accordance with the modernized metric system known as the International System of Units (SI). Much of the world utilizes the metric system, and because of this, it is necessary that the *NEC* accommodates the SI system.

90.9(B) Dual System of Units - Both inch-pound and metric units are referenced together when the *NEC* requires a specific distance, size, or weight. During the code-making cycle which produced the 2002 edition of the *NEC*, metric units were recognized as the international standard and all metric units were placed first, with the inch-pound units following in parentheses. For example: A cable assembly or other wiring method may require support every 1.8 m (6 ft.) and within 300 mm (12 in.) of all terminations.

90.9(C) Permitted Uses of Soft Conversions - 90.0(C) instructs users on how conversions from inch-pound to metric and vice versa are permitted to be made. An exact conversion from inch-pound to metric is known as a soft conversion. When a requirement within the *NEC* directly impacts safety, such as working space in **110.26(A)**, a soft, or exact, conversion is required. A conversion that is permitted to be approximate or rounded off to the nearest standard unit is called a hard conversion. For standardization of requirements for support of wiring methods, hard or rounded-off conversions are used. Four cases of soft conversions are given in the *Code*:

90.9(C)(1) Trade Sizes
90.9(C)(2) Extracted Material
90.9(C)(3) Industry Practice
90.9(C)(4) Safety

90.9(D) Compliance - Approximate, or hard, conversions that do not negatively impact safety are permitted by the *Code*. Compliance with either the metric or the inch-pound system meets the requirements of the *NEC*. The choice of which units to apply, metric or inch-pound, is determined by the *Code* user.

NOTATIONS USED IN THE *NEC*

Notations appear in the *Code* to help the user identify the changes from the previous edition if they are not explained in **Article 90**. This editorial information is handled directly by NFPA staff and is provided on the inside cover of the *NEC*.

These notations are as follows:

1. A vertical line is placed next to a technical change from the previous edition where any of the following occurs:
   - The change occurs in large blocks
   - Large areas of new text are added
   - The change occurs in tables and figures
   - New tables or figures are added
2. Changes other than editorial are highlighted with gray shading within sections.
3. Where one or more complete paragraphs are deleted, a bullet (•) is placed in the margin between the remaining paragraphs or sections. **See Figure 6-12.**

Figure 6-12	Paragraph Deletion

110.19 Light and Power from Railway Conductors. Circuits for lighting and power shall not be connected to any system that contains trolley wires with a ground return.

Exception: Such circuit connections shall be permitted in car houses, power houses, or passenger and freight stations operated in connection with electric railways.

•

110.21 Marking. The manufacturer's name, trademark, or other descriptive marking by which the organization responsible for the product can be identified shall be placed on all electrical equipment. Other markings that indicate voltage, current, wattage, or other ratings shall be provided as specified elsewhere in this *Code*. The marking shall be of sufficient durability to withstand the environment involved.

Reprinted with permission from NFPA 70-2011, *National Electrical Code*®, Copyright© 2010, National Fire Protection Association, Quincy, MA 02169. This reprinted material is not the complete and official position of the NFPA on the referenced subject, which is represented only by the standard in its entirety.

Figure 6-12. The bullet indicates that 110.20 was removed in the 2011 NEC.

Summary

The nine sections comprising **Article 90** provide the ground rules for the use, application, and scope of the entire *National Electrical Code*. The nine sections contained in this article are as follows:

- Section 90.1 Purpose
- Section 90.2 Scope
- Section 90.3 Code Arrangement
- Section 90.4 Enforcement
- Section 90.5 Mandatory Rules, Permissive Rules, and Explanatory Material
- Section 90.6 Formal Interpretations
- Section 90.7 Examination of Equipment for Safety
- Section 90.8 Wiring Planning
- Section 90.9 Units of Measurement

The *NEC* Introduction must be understood prior to any use of the rules or information contained in the *Code*. These directions are given to the users of the *Code* for proper understanding and application of the *NEC*.

Utility transmission transformer in Utility Substations are not covered in the NEC.

Review Questions

1. The primary purpose of the *National Electrical Code* is __?__.
   a. cost-efficient installations
   b. to increase installation costs
   c. safety
   d. training

2. The *NEC* contains __?__ for electrical installations.
   a. provisions for convenient expansion
   b. provisions for training of persons
   c. a design manual
   d. minimum requirements

3. The *NEC* is not intended for use as __?__.
   a. an instruction manual
   b. a tool for untrained persons
   c. a design manual
   d. all of the above

4. The use of mandatory rules, permissive rules, and explanatory material throughout the *NEC* is governed by Article 90 in Section __?__.
   a. 90.1
   b. 90.3
   c. 90.2
   d. 90.5

5. The use of the term "shall" represents __?__ text in the *NEC*.
   a. mandatory
   b. permissive
   c. informational
   d. none of the above

6. The use of the phrase "shall be permitted" represents __?__ text in the *NEC*.
   a. mandatory
   b. optional
   c. informational
   d. none of the above

7. The use of Informational Notes represents __?__ text in the *NEC*.
   a. mandatory
   b. optional
   c. informational
   d. none of the above

8. The *Code* user is permitted by __?__ to use metric or inch-pound units to be in compliance with the *NEC*.
   a. 90.5(B)
   b. 90.2(C)
   c. 90.9(B)
   d. 90.9(D)

Chapter 1 of the *NEC*, "GENERAL"

The *Codeology* title for Chapter 1 is "General." All electrical installations are subject to the general rules which apply in all electrical installations. For example, the definitions given in **Article 100** apply to the terms wherever they are used in the *Code*. General requirements also exist in the "General" chapter, **Article 110**, covering basic needs which are common to all electrical installations. In accordance with the *Code* arrangement requirements of **Section 90.3**, Chapter 1 applies to all electrical installations unless supplemented or modified by Chapters 5, 6, or 7.

Objectives

» Label the *Codeology* title for *NEC* Chapter 1 as "General."

» Summarize the general type of information and requirements contained in Chapter 1.

» Identify Chapter 1 numbering as the 100-series.

» Recognize, recall, and apply the articles contained in Chapter 1.

Chapter 7

Table of Contents

NEC CHAPTER 1 GENERAL

Chapter 1, the 100-series, is comprised of two articles, one providing the definitions essential for the proper application of the *NEC* (**Article 100**) and the second providing the general requirements for all electrical installations (**Article 110**). As will be discussed later, **Article 110** is broken down into five parts.

Article 100 contains only those definitions essential to the proper application of this *Code*. It is not intended to include commonly defined general terms or commonly defined technical terms from related codes and standards. In general, only those terms which are used in two or more articles are defined in **Article 100**. Other definitions are included in the article in which they are used, but may be referenced in **Article 100**. **Figure 7-1** reviews the articles for Chapter 1.

Article 100 contains definitions for terms which are not commonly used or are essential to the proper application of the *NEC*. A term which needs to be defined in the *NEC* is placed in **Article 100** only if it is used in two or more articles. Terms which are defined in the *NEC* and are used in only one article are defined in the second section of the article in which they appear. For example, in **Article 517 Health Care Facilities**, forty definitions are listed in **Section 517.2**, the second section of the article. When terms are defined within an individual article, they apply only within that article.

Part I of **Article 100** contains definitions intended to apply wherever the terms are used throughout the *Code*. **Part II** contains four definitions applicable only to the parts of articles specifically covering installations and equipment operating at over 600 volts, nominal. These four definitions are:

1. Electronically Actuated Fuse
2. Fuse
3. Multiple Fuse
4. Switching Device

For example, terms defined in **Part II** of **Article 100** would apply to **Part IX** of **Article 240**. **See Figure 7-1.**

Misapplication of the *Code* requirements occurs regularly due to a lack of understanding of terms defined in **Article 100**. For instance, to understand and apply the requirements of **Article 230 Services**, the user must understand the terms defined in **Article 100**, including but not limited to:

Service. The conductors and equipment for delivering electric energy from the serving utility to the wiring system of the premises served.
Service Cable. Service conductors made up in the form of a cable.
Service Conductors. The conductors from the service point to the service disconnecting means.

Figure 7-1	Layout of *NEC*® Chapter 1

NEC Title: General
Codeology Title: General
Chapter Scope: General Information and Rules for Electrical Installations

Article	Title
100	Definitions
110	Requirements for Electrical Installations

Figure 7-1. Article 100 and Article 110 of Chapter 1 applies to all electrical installations unless altered by Code language of Chapters 5, 6, and 7.

Service Conductors, Overhead. The overhead conductors between the service point and the first point of connection to the service-entrance conductors at the building or other structure.

Service Conductors, Underground. The underground conductors between the service point and the first point of connection to the service-entrance conductors in a terminal box, meter, or other enclosure, inside or outside the building wall.

Informational Note: Where there is no terminal box, meter, or other enclosure, the point of connection is considered to be the point of entrance of the service conductors into the building.

Service Drop. The overhead conductors between the utility electric supply system and the service point.

Service-Entrance Conductors, Overhead System. The service conductors between the terminals of the service equipment and a point usually outside the building, clear of building walls, where joined by tap or splice to the service drop or overhead service conductors. **See Figure 7-2.**

Service-Entrance Conductors, Underground System. The service conductors between the terminals of the service equipment and the point of connection to the service lateral or underground service conductors. **See Figure 7-3.**

Informational Note: Where service equipment is located outside the building walls, there may be no service-entrance conductors or they may be entirely outside the building.

Service Equipment. The necessary equipment, usually consisting of a circuit breaker(s) or switch(es) and fuse(s) and their accessories, connected to the load end of service conductors to a building or other structure, or an otherwise designated area, and intended to constitute the main control and cutoff of the supply.

Service Lateral. The underground conductors between the utility electric supply system and the service point.

Service Point. The point of connection between the facilities of the serving utility and the premises wiring.

Informational Note: The service point can be described as the point of demarcation between where the serving utility ends and the premises wiring begins. The serving utility generally specifies the location of the service point based on the conditions of service.

The term "service" is actually part of various definitions of **Article 100** other than the ones above: Bonding Jumper, Main; Feeder; Listed; Overcurrent Protective Device, Branch-Circuit; Premises Wiring (System); Separately Derived System; and Surge-Protective Device (SPD). **Article 100** contains many additional definitions which are essential to the proper application of **Article 230**. However, nine separate definitions are needed to properly identify conductors and equipment in order to ensure proper application of **Article 230**. Before interpreting and applying the rules of **Article 230** or any other article, the *Code* user must understand the definitions which apply.

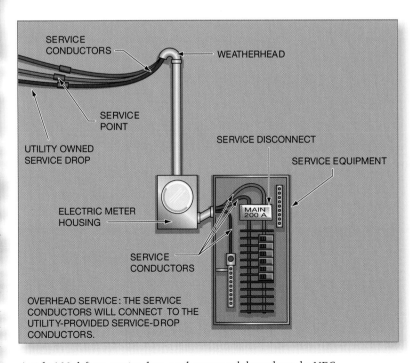

Article 100 defines service drop conductors used throughout the NEC.

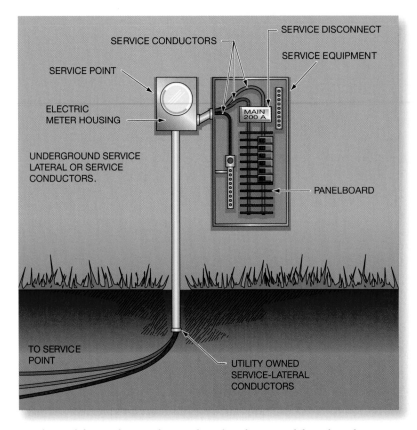

SERVICE CONDUCTORS

SERVICE POINT

SERVICE DISCONNECT

SERVICE EQUIPMENT

ELECTRIC
METER HOUSING

UNDERGROUND SERVICE
LATERAL OR SERVICE
CONDUCTORS.

MAIN
200 A

PANELBOARD

TO SERVICE
POINT

UTILITY OWNED
SERVICE-LATERAL
CONDUCTORS

Article 100 defines underground service lateral conductors used throughout the NEC.

ARTICLE 110 REQUIREMENTS FOR ELECTRICAL INSTALLATIONS

Article 110 contains general requirements, separated into five parts, which are common to all electrical installations. **See Figure 7-2.** The following key provisions of this article represent the general nature of these requirements as they apply to all electrical installations.

I. General

Section 110.1 Scope

These general requirements for all electrical installations include but are not limited to the following:

- Examination and approval of conductors and equipment
- Installation and use of conductors and equipment
- Access to electrical equipment
- Spaces about electrical equipment
- Enclosures intended for personnel entry
- Tunnel installations

Besides the common requirements listed in this section, a complete review of all the *Code* language of **Article 110** is very important.

Section 110.2 Approval

This section requires that all conductors and equipment be approved as defined in **Article 100**. Approval occurs when the conductors and equipment are acceptable to the authority having jurisdiction (AHJ).

> Informational Note: See 90.7, Examination of Equipment for Safety, and 110.3, Examination, Identification, Installation, and Use of Equipment. See definitions of Approved, Identified, Labeled, and Listed.

Section 110.3 Examination, Identification, Installation, and Use of Equipment

Figure 7-2	Five Parts of Article 110
Part I.	General
Part II.	600 Volts, Nominal, or Less
Part III.	Over 600 Volts, Nominal
Part IV.	Tunnel Installations over 600 Volts, Nominal
Part V.	Manholes and Other Electric Enclosures Intended for Personnel Entry, All Voltages

Figure 7-2. Article 110 is divided into five parts which identify the basic requirements for electrical installations.

110.3(A) Examination - This *Code* language requires that the equipment be examined to evaluate:

1. Suitability
2. Mechanical strength and durability
3. Wire-bending and connection space
4. Electrical insulation
5. Heating effects
6. Arcing effects
7. Classification by type, size, voltage, current capacity, and use
8. Other factors that contribute to the practical safeguarding of persons

110.3(B) Installation and Use - Listed or labeled equipment is required to be installed in accordance with the listing and labeling of the equipment. **See Figure 7-3.** For example, a conductor termination point labeled CU is permitted only for copper conductors while the labeling of AL/CU on a termination point permits copper or aluminum conductors.

Section 110.5 Conductors

Conductors referenced in the *NEC* shall be copper unless otherwise provided in the *Code*. When the conductor material is not specified in a *Code* section, the material and sizes given apply to copper conductors. Conductor sizes and ampacity for aluminum and copper-clad aluminum conductors are listed and used on several *Code* tables.

Informational Note: For aluminum and copper-clad aluminum conductors, see 310.15.

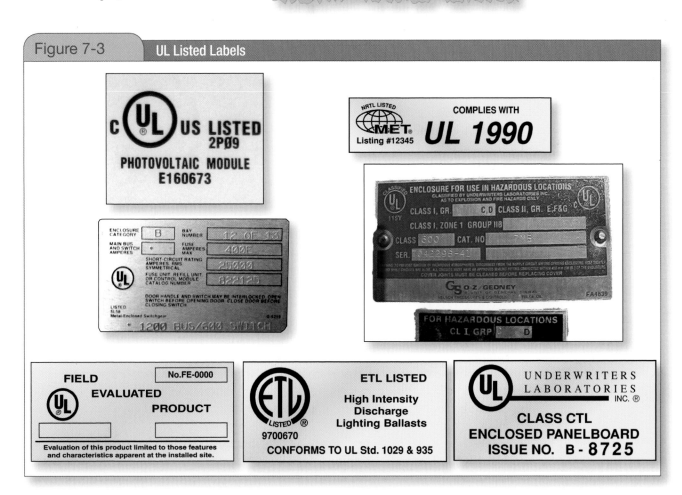

Figure 7-3   UL Listed Labels

Figure 7-3. Underwriters Laboratories' first test was in 1895 under the name of Underwriters' Electrical Bureau. Today, the electrical equipment installed is listed in the UL White Book.

Section 110.6 Conductor Sizes

Conductor sizes used in the *NEC* are expressed in American Wire Gage (AWG) or in circular mils. Table 8 in Chapter 9 shows the circular mil sizes for conductors using the AWG. Examples of AWG sizing is #12 AWG and #1/0 AWG. Circular mils are typically in kcmil (1,000 mils). An example is 500 kcmil. Several *Code* cycles back, the kcmils were once referred to as MCM. Older electricians may still use this reference.

Section 110.9 Interrupting Rating

All equipment intended to interrupt current at fault levels must have a sufficient interrupting rating.

Section 110.10 Circuit Impedance, Short-Circuit Current Ratings, and Other Characteristics

Damage to electrical systems from short circuits and ground faults must be minimized through proper selection of protective equipment and consideration of all characteristics of the equipment and conductors protected.

Section 110.11 Deteriorating Agents

Equipment that is not identified for use outdoors or identified for indoor use shall be protected from damage from the weather during construction. Plastic sheeting or other means for protection is commonly used to prevent this damage.

Section 110.12 Mechanical Execution of Work

This section requires that all electrical equipment be installed in a neat and workmanlike manner. Electrical installations must be installed neatly and with care to be considered installed in a skillful or workmanlike manner.

110.12(A) Unused Openings - All unused cable or raceway openings (holes) in all electrical equipment must be effectively closed. Holes intended for the operation of equipment or mounting purposes do not have to be closed.

110.12(B) Integrity of Electrical Equipment and Connections - All electrical equipment must be kept clean and undamaged by foreign materials such as paint, plaster, or other corrosive materials. Equipment must be protected from damage, corrosion, overheating, or deteriortation.

Section 110.13 Mounting and Cooling of Equipment

110.13(A) Mounting - All electrical equipment must be firmly secured to the surface on which it is mounted. Wooden plugs are prohibited for mounting or securing of electrical equipment.

Q. Why is this?

A. In the event of a fire, wooden plugs fail quickly and the equipment may fall causing injury to those exiting from the location.

110.13(B) Cooling - Electrical equipment that depends on the natural circulation of air for cooling purposes must be installed such that other equipment, walls, or other structural members do not block or limit the natural circulation of air. Because electrical equipment, such as a transformer, generates heat during normal operation, this *Code* language requires that all electrical equipment which might generate heat during normal operation be installed to allow for cooling through the natural circulation of air.

Section 110.14 Electrical Connections

All electrical installations contain electrical connections. This section contains general requirements for all electrical connections in all electrical installations. These requirements are as follows:

- Due to the different characteristics of conductor metals, such as aluminum and copper, all terminals, connectors, and lugs shall be identified (labeled/marked) for the conductor material permitted.
- Dissimilar metals, such as aluminum and copper, shall not be intermixed in a connector unless the connector is identified or listed for the purpose.
- Solder, flux, inhibitors, and conductor compounds must not have an adverse effect on the conductors or equipment.

> Informational Note: Many terminations and equipment are marked with a tightening torque.

110.14(A) Terminals - Conductors must be terminated in such a manner to ensure a solid connection without damaging the conductors. Terminals designed for the termination of more than one conductor must be identified for such use. Terminals designed for use with aluminum conductors must also be identified for the purpose.

110.14(B) Splices - Conductors are required to be spliced using one of the following methods:

- Devices identified for the purpose (for example, wire nuts)
- Brazing, welding, or soldering with a fusible metal or alloy
- Soldered splices must first be mechanically joined
- All splices and free ends of conductors must be insulated
- Wire nuts or splicing devices for direct burial must be listed for the purpose

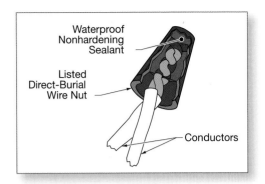

For additional information, visit qr.njatcdb.org Item #1065

For smaller conductors, the wire nut is the most common method of splicing.

110.14(C) Temperature Limitations - All electrical terminations and conductors are temperature limited, typically at 60°C, 75°C, and 90°C. The intent of this requirement is to ensure that the lowest temperature rating in an electrical system or circuit is not exceeded. For example, a circuit may be fed from an overcurrent device rated at 75°C using conductors rated at 90°C and terminate in equipment rated at 60°C. The lowest temperature rating of 60°C must then be applied. Derating of the conductor, if required, is permitted to start at the higher 90°C rating.

Courtesy of Fluke

The termination temperature is measured by a Fluke Infrared Thermometer.

Section 110.15 High-Leg Marking

The use of a 3-phase, 4-wire delta connected system results in one phase having a higher voltage to ground. For example, a 120/240 volt, 3-phase, 4-wire system is required, by **Article 250** of the *NEC*, to be grounded. The voltage to ground from "A" phase/leg is 120 volts. The voltage to ground from "C" phase/leg is 120 volts. The "B" phase/leg voltage to ground is significantly higher at 208 volts to ground. This section requires that the high leg be identified with the color orange or other effective means wherever connections are made and both the grounded conductor and the high leg are available. This marking is required to prevent misapplication of the high leg. **See Figure 7-4.**

Section 110.16 Arc-Flash Hazard Warning

Electrical hazards include shock hazards, arc flash hazards, and arc blast hazards. An arc flash can produce temperatures of 35,000°F at the point of contact, and temperatures in the ambient space of an electrical worker can easily reach upward of 15,000°F. The result of an accident involving an arc flash can result in serious injuries from incurable third-degree burns and death.

The marking requirement in **Section 110.16** is designed to warn qualified persons of potential arc flash hazards. **See Figure 7–5.** Arc flash warnings are required to be marked on all switchboards, panelboards, industrial control panels, meter socket enclosures, and motor control centers in other than dwelling units.

An informational note informs the *Code* user that *NFPA-70E Standard for Electrical Safety in the Workplace* provides guidance in determining the severity of an exposure, safe work practices, and required personal protective equipment.

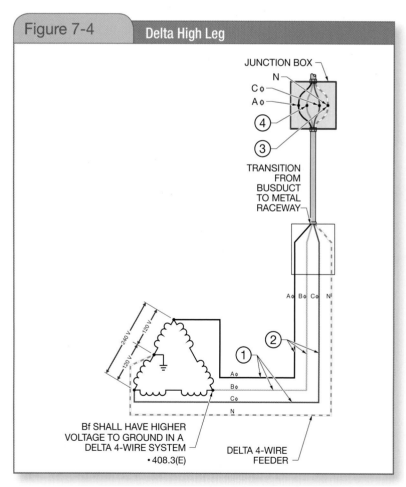

Figure 7-4. *The high leg, often called the wild leg, must be installed as the "B" phase on a 4-wire delta system and identified with the color orange.*

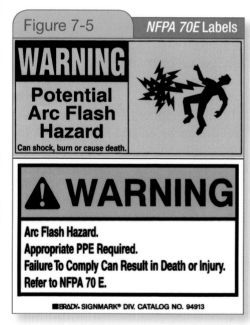

Figure 7-5. *Personal protection equipment is available in various calorie ratings. NFPA 70E is a valuable resource to be studied for electrical worker arc flash protection.*

Section 110.21 Marking

All electrical equipment is required to be durably marked with the manufacturer's name, trademark, or other distinctive marking for identification purposes.

Section 110.22 Identification of Disconnecting Means

All disconnecting means are required to be legibly marked to identify their purpose, unless the installation is so arranged that the purpose of the disconnecting means is clearly evident. Other markings required include voltage, current, wattage, or other ratings required elsewhere in this *Code*.

Section 110.24 Available Fault Current

Service equipment, other than dwelling units, shall be marked legibly in the field listing the maximum available fault current. This marking shall include the date the fault current was calculated. Upstream equipment such as transformer size, distance to the next distribution equipment or transformer, motor loads, and other factors are all part of the calculation for available fault current.

II. 600 Volts, Nominal, or Less

Section 110.26 Spaces About Electrical Equipment

This section requires that sufficient access to equipment and working space be maintained.

All electrical equipment likely to require examination, adjustment, servicing, or maintenance while energized must provide adequate working space as required in **110.26(A)(1), (2), and (3)**.

110.26(A) Working Space

(1) Depth of working space. Adequate working space is required for the safety of electrical workers who install and maintain electrical systems. **Table 110.26(A)(1) Working Spaces** specifies the minimum working space in front of electrical equipment. This table classifies

systems 600 volts or less in either of two categories using the voltage to ground system: 0 to 150 volts to ground and 151 to 600 volts to ground. Minimum clear distance for working space is then determined from one of three conditions:

Condition 1 – The equipment is opposite no other electrical equipment or any grounded objects or parts.

Condition 2 – The equipment is opposite grounded objects or parts. Note that concrete, brick, or tile walls are considered grounded.

Condition 3 – The equipment is opposite other electrical equipment.

(2) Width of working space. The width of working space in front of all electrical equipment is required to be the width of the equipment or 30 inches, whichever is greater. In all cases, hinged panels and doors must be capable of opening a full 90 degrees.

(3) Height of working space. The working space must be clear and extend from the floor or working platform to the height of the equipment or 6 ½ feet, whichever is greater. **See Figure 7–6.**

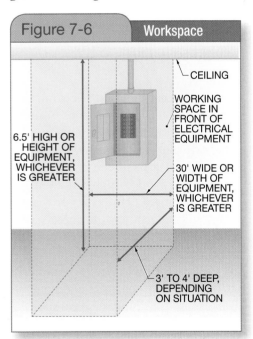

| Figure 7-6 | Workspace |

CEILING

WORKING SPACE IN FRONT OF ELECTRICAL EQUIPMENT

6.5' HIGH OR HEIGHT OF EQUIPMENT, WHICHEVER IS GREATER

30' WIDE OR WIDTH OF EQUIPMENT, WHICHEVER IS GREATER

3' TO 4' DEEP, DEPENDING ON SITUATION

Figure 7-6. Nothing can be stored in the work space area so that personnel can access the electrical equipment quickly.

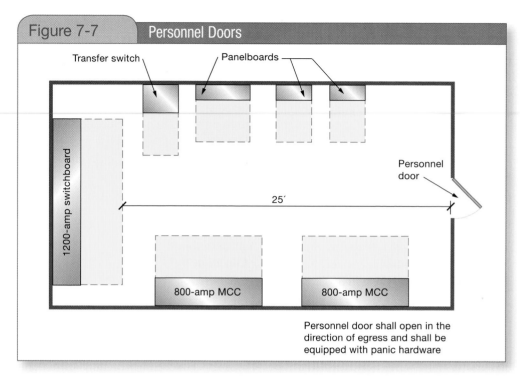

Figure 7-7 | **Personnel Doors**

Transfer switch

Panelboards

1200-amp switchboard

Personnel door

25′

800-amp MCC

800-amp MCC

Personnel door shall open in the direction of egress and shall be equipped with panic hardware

Figure 7-7. The NEC requires panic hardware on doors for large electrical rooms.

110.26(B) Clear Spaces - The working space in front of electrical equipment must not be used for storage.

110.26(C) Entrance to and Egress from Working Space

(1) Minimum required. In general, a minimum of one entrance is required for access to working space about electrical equipment.

(2) Large equipment. In general, for equipment rated at 1,200 amps or more, over 6 feet wide, and containing overcurrent, switching, or control devices, one entrance at each end of the working space not less than 6 ½ feet high and 24 inches wide is required. However, provisions permit a single entrance to these spaces where the exit is unobstructed or additional working space is provided.

(3) Personnel Doors. Where equipment rated 1,200 amps or more and contains overcurrent, switching, or control devices, all personnel doors less than 25 feet from the nearest edge of the working space must open in the direction of egress and be equipped with panic-type hard-

ware to permit a quick exit in the event of an electrical fault. **See Figure 7–7.**

110.26(D) Illumination - Working spaces around service equipment, switchboards, panelboards, or motor control centers installed indoors require illumination to provide electrical workers maintaining the system with adequate lighting to perform routine tasks.

110.26(E)(1)(a) Dedicated Electrical Space - This section applies to all switchboards, panelboards, and motor control centers and requires that where they are installed, the locations must be dedicated to this equipment. This requirement is not working space for the electrical worker; rather it is dedicated equipment space to allow for raceways and cable assemblies to enter equipment. This requirement is also intended to prevent foreign systems from being installed above electrical equipment. The dedicated space addressed in this section is the area above the footprint formed by the top of the equipment. **See Figure 7-8.** In general, the area 6 feet above the footprint formed by the top of the equipment or the structural ceiling, whichever is lower, is dedicated equipment space.

III. Over 600 Volts, Nominal

Part III of **Article 110** is dedicated to general requirements for all electrical installations for systems rated at over 600 volts. **Section 110.30** clearly explains that the requirements of **Part III** are in addition to the general provisions of **Part I** of **Article 110**. This section also emphasizes that **Part III** does not apply on the supply side or upstream of the service point. The general requirements in **Part III** address areas very similar to those covered in **Part II** of **Article 110**. The difference is that the requirements of **Part III** are modified to address the safety of high-voltage system installations.

IV. Tunnel Installations over 600 Volts, Nominal

Part IV of **Article 110** is dedicated to tunnel installations with systems and circuits rated over 600 volts. **Section 110.51** explains in detail the areas covered by **Part IV.** This part covers the installation and use of high-voltage power distribution and utilization equipment which is portable, mobile, or both. Examples include, but are not limited to, substations, trailers, cars, mobile shovels, draglines, hoists, drills, dredges, compressors, pumps, conveyors, underground excavators, and the like.

V. Manholes and Other Electrical Enclosures Intended for Personnel Entry, All Voltages

Part V of **Article 110** is dedicated to manholes and other electric enclosures intended for personnel entry at all voltages. It addresses working space inside these enclosures and the provision of adequate access and space where equipment or parts contained are likely to require examination, adjustment, servicing, or maintenance while energized. Enclosures must also be

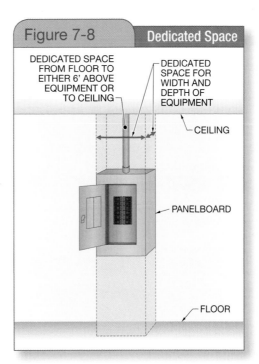

Figure 7-8   **Dedicated Space**

DEDICATED SPACE FROM FLOOR TO EITHER 6' ABOVE EQUIPMENT OR TO CEILING

DEDICATED SPACE FOR WIDTH AND DEPTH OF EQUIPMENT

CEILING

PANELBOARD

FLOOR

Figure 7-8. No piping, ducts, leak protection apparatus, or other equipment foreign to the electrical installation shall be located in the dedicated space.

of sufficient size to allow for the installation and removal of conductors without damaging the conductor insulation.

KEY WORDS AND CLUES FOR CHAPTER 1, "GENERAL"

- General definitions for terms used in more than one article
- Determination of approval
- Examination of equipment
- Identification of equipment
- Installation and use of equipment
- General questions about electrical installations
- Mechanical execution of work
- Mounting and cooling of equipment
- Electrical connections
- High-leg marking requirements
- Flash protection
- Markings
- General identification of disconnects
- General electrical questions for over 600 volts
- General questions for manholes and tunnels
- Working space

Summary

In accordance with **Section 90.3**, Chapter 1 applies generally to all electrical installations. Using the *Codeology* method, this chapter has been coined the "General" chapter, due to the breadth of its coverage. The scope of this chapter, which can be described as "General Information and Rules for all Electrical Installations," covers the entire electrical system, from the service point (connection to the utility) to the last receptacle or outlet in the electrical system.

Chapter 1, along with Chapters 2, 3, and 4, builds the foundation or backbone of all electrical installations. **See Figure 7-9.**

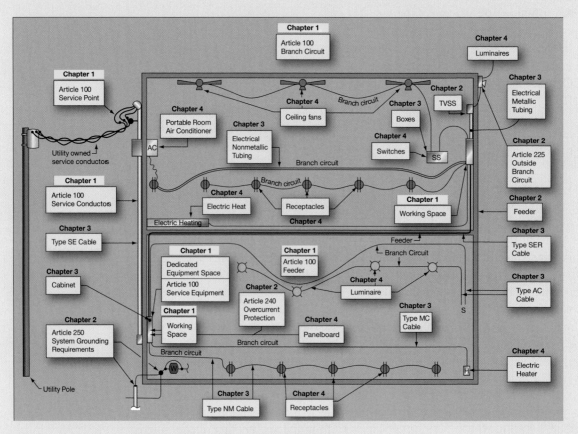

Figure 7-9. Chapter 1 applies generally in all electrical installations.

Review Questions

1. Does Chapter 1 of the *NEC* apply to all electrical installations covered by the *Code*?

2. Chapter 1 is divided into how many articles?

3. Does Article 100 Definitions contain all terms defined in the *NEC*?

4. When is a definition placed in Article 100 of the *NEC*?

5. When a term is defined in an article other than Article 100, it is always placed in which section?

6. Do the provisions of Part II of Article 100 apply to the installation of a 120/240 volt panelboard in a dwelling unit?

7. Workspace clearances are general requirements for all electrical installations. Where in Chapter 1 is this general requirement for equipment rated at 240 volts located?

8. Article 110 is divided into how many parts?

9. Is Part IV of Article 110 limited to installations above 600 volts?

10. Which part of Article 110 would contain the requirement that all listed and labeled equipment be installed in accordance with their listing and labeling?

To access Practice Problems, visit qr.njatcdb.org. Item #1082 Click on Inside Blended Learning.

Chapter 2 of the *NEC*, "PLAN"

The *Codeology* title for Chapter 2 is "Plan." All electrical installations must be planned to suit many needs while taking into account the total cost of materials and installation.

The minimum needs include:

- **Design:** electrical installation must suit the specific needs of the occupancy, building, structure, or location in which it is installed.
- **Compliancy:** installation practices must fall within the *NEC* and the local building codes requirements
- **Plan submission:** Conceptual and final drawings must be submitted to the authority having jurisdiction (AHJ).
- **Planning Stages:** The planning stage of all electrical installations includes two very important areas—wiring and protection.

Objectives

» Associate the *Codeology* title for *NEC* Chapter 2 as "Plan."

» Identify the planning type of information and requirements for wiring and protection contained in Chapter 2.

» Recognize Chapter 2 numbering as the 200-series.

» Recognize, recall, and apply the articles contained in Chapter 2.

Chapter 8

Table of Contents

WIRING

In the planning stages of any electrical installation, the designer's intent can be seen on the electrical drawings. These drawings provide the installer with the necessary information to build the installation. All electrical installations must have a source of electrical energy, that is, in most cases, a service (**Article 230**) from a local utility company. From the service equipment, feeders (**Article 215**) supply panelboards to distribute electrical energy throughout the building. From the panelboards, branch circuits (**Article 210**) supply receptacle outlets, lighting fixtures, and other utilization equipment to facilitate the use of electrical energy. Branch circuits and feeders installed outside of a building or structure must also comply with outside branch circuits and feeders (**Article 225**). Additional articles within Chapter 2 are dedicated to wiring. **See Figure 8-1**.

All of the different types of conductors must be properly sized to handle the intended load. This requires that calculations (**Article 220**) be applied to properly size each conductor. When the electrical system employs a grounded conductor (**Article 200**), the installer must plan for the proper use and identification of these conductors. **See Figure 8-2**.

Code compliance for current-carrying conductors must be part of the planning process of the electrical installation. This process, covered in the wiring articles of Chapter 2, focuses only on the *NEC* terms for the current-carrying conductors, not the type of raceway or cable assembly. **See Figure 8-3**. These *NEC* terms for conductors are major clues for starting in Chapter 2. *NEC* terms related to planning the wiring of an electrical installation include the following:

- Branch circuits
- Feeders
- Services
- Calculations, computed load(s)
- Grounded conductors

210.19(A)(1) Informative Note No. 4 requires the increase of conductor size for long conductor runs to prevent circuit voltage drop. Long conductor runs increase the circuit's resistance, which in turn reduces the voltage at the equipment being served. The *NEC* allows for up to 3% voltage drop for general branch circuits. A 208-volt source with a 3% voltage drop due to a long conductor run will only provide 202 volts at the equipment being served. Voltage drops greater than 3% would likely damage the equipment being served or cause it to perform poorly. Although it is beyond the scope of this text, a quick/rough calculation for determining the conductor size (CM) to accommodate voltage drop in a long run is:

$$CM = \frac{Ph \times L \times I \times K}{E \times \%VD}$$

Where:
CM: Circular Mils (see *NEC* Table 8)
Ph: 1.732 for 3-ph; 2 for 1-ph
L: One-way length
K: 12.9 for copper; 21 for alum
E: Nominal voltage
%VD: 0.03 for 3% branch VD

Figure 8-1	Chapter 2 Wiring (and Protection)	
NEC® Title:	Wiring and Protection	
Codeology Title:	Plan	
Chapter Scope:	Information and Rules on *Wiring and Protection* of Electrical Installations	
WIRING		
Article	Article Title	
200	Use and Identification of Grounded Conductors	
210	Branch Circuits	
215	Feeders	
220	Branch Circuit, Feeder, and Service Calculations	
225	Outside Branch Circuits and Feeders	
230	Services	

Figure 8-1. Articles 200, 210, 215, 220, 225, and 230 cover the wiring of Chapter 2.

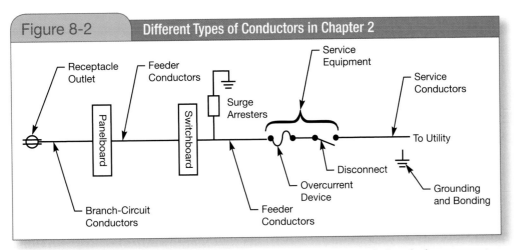

Figure 8-2. Chapter 2, the "Plan" chapter, does not address specific wiring methods, only the terms service, feeder, and branch circuit.

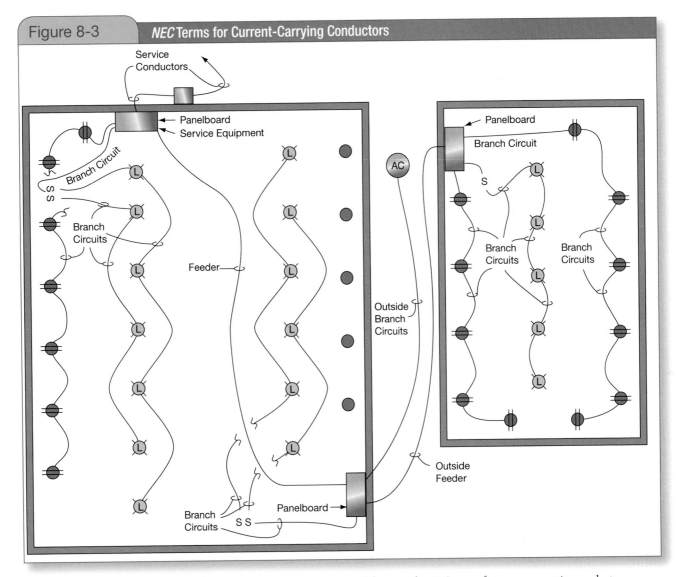

Figure 8-3. The planning requirements in the wiring articles of Chapter 2 focus on the NEC terms for current-carrying conductors.

Article 200 Use and Identification of Grounded Conductors

The term *grounded conductor* is defined in **Article 100. See Figure 8-4.**

> **Grounded Conductor.** A system or circuit conductor that is intentionally grounded.

Grounded conductors are often called the neutral conductor. It is extremely important to note that not all grounded conductors are neutrals. When a current-carrying conductor is common to all of the ungrounded (hot) conductors of the system, it is considered neutral. When that common or neutral conductor is intentionally grounded, it becomes the grounded conductor.

Article 200 provides requirements for the following:

1. Identification of terminals for grounded conductors

2. Use of grounded conductors in premises wiring systems
3. Identification of grounded conductors

Article 210 Branch Circuits

The term *branch circuit* is defined in **Article 100. See Figure 8-5.**

> **Branch Circuit.** The circuit conductors between the final overcurrent device protecting the circuit and the outlet(s).

This basic definition defines the current-carrying conductors which supply current to the load. The different types of branch circuits are defined in **Article 100** to differentiate between the different uses of branch circuits and the specific rules within the *NEC* for each type. **Article 210** provides the general requirements for all branch circuits. Note that in the scope of this article, branch circuits for motor loads are covered in **Article 430. Section 210.2** provides a cross-reference table to guide the *Code* user to the correct article for specific-purpose branch circuits.

Article 210 is subdivided into three logical parts as follows:

I. General Provisions
- Ratings
- Identification
- Voltage Limitations
- Receptacle Requirements
- GFCI Requirements
- AFCI Requirements
- Required Branch Circuits

II. Branch-Circuit Ratings
- Conductor Ampacity and Size
- Overcurrent Protection
- Outlet Devices
- Permissible Loads
- Common Area Branch Circuits

III. Required Outlets
- Dwelling Units
- Guest Rooms
- Show Windows
- Lighting Required
- HVAC Outlet Required

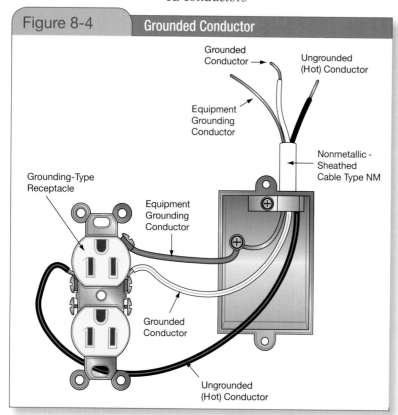

Figure 8-4   **Grounded Conductor**

Grounded Conductor →

Ungrounded (Hot) Conductor

Equipment Grounding Conductor

Nonmetallic - Sheathed Cable Type NM

Grounding-Type Receptacle

Equipment Grounding Conductor

Grounded Conductor

Ungrounded (Hot) Conductor

Figure 8-4. The NEC defines all current-carrying and grounding conductors. The "ungrounded" conductor is commonly called the hot conductor. The "grounded" conductor is commonly called the neutral.

Article 215 Feeders

Article 100 defines the term *feeder*. **See Figure 8-6.**

Feeder. All circuit conductors between the service equipment, the source of a separately derived system, or other power supply source and the final branch-circuit overcurrent device.

The three primary types of conductors addressed in the *NEC* are service, feeder, and branch circuit. Service conductors may originate only at a utility owned and supplied source, and end where disconnecting means and overcurrent protection are provided. Branch circuits begin at the final overcurrent protective device and end at the outlet or utilization equipment supplied. Feeders are, very simply, all of the conductors in between the service equipment and the final overcurrent protective device. A hierarchy does not exist for feeders. The term *feeder* is often misused as a "subfeeder," a term which does not exist in the *NEC*. Therefore, there is no such thing as a "subfeeder."

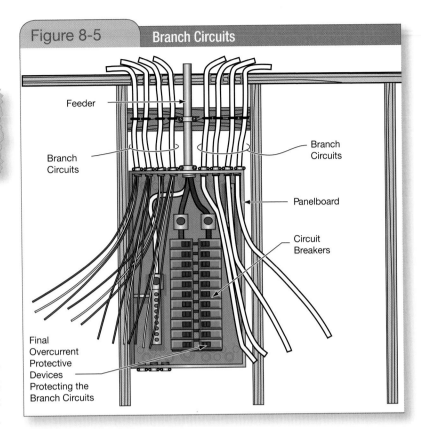

Figure 8-5. The conductors from the final overcurrent device, fuse, or circuit breaker to the receptacles, lighting outlets, hardwired equipment, and all other outlets are branch-circuit conductors.

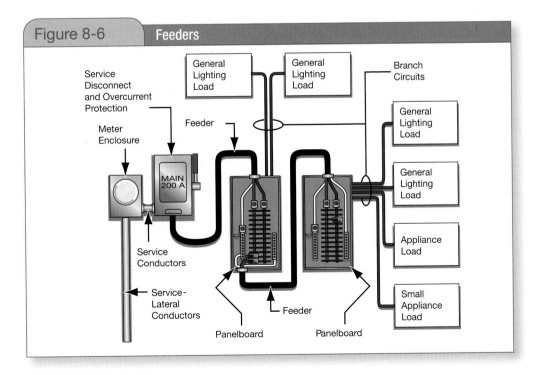

Figure 8-6. All conductors between the final overcurrent protective device and the service, or other power supply, are feeder conductors.

Article 215 provides requirements for the following:

1. Installation of feeders
2. Overcurrent protection for feeders
3. Minimum size and ampacity of feeders

Article 220 Branch-Circuit, Feeder, and Service Calculations

Article 220 is subdivided into five logical parts as follows:

I. General
II. Branch-Circuit Load Calculations
III. Feeder and Service Load Calculations
IV. Optional Feeder and Service Load Calculations
V. Farm Load Calculation

Article 220 provides requirements for the following:

1. Computing branch-circuit loads
2. Computing feeder loads
3. Computing service loads

Article 225 Outside Branch Circuits and Feeders

While **Article 210** and **Article 215** provide the basic requirements for branch circuits and feeders, **Article 225** provides specific requirements for outside branch circuits and feeders as follows:

1. Those run on or between buildings
2. Those on structures
3. Those on poles on the premises
4. Those supplying electric equipment and wiring for the supply of utilization equipment located on or attached to the outside of buildings, structures, or poles. **See Figure 8-7.**

Section 225.3 provides a cross-reference table to guide the *Code* user to the correct article where the *NEC* provides additional requirements for outside branch circuits and feeders.

Figure 8-7 — Outside Circuit

Figure 8-7. Article 225 applies to all feeders and branch circuits located outside.

Article 225 is subdivided into three logical parts as follows:

I. General
This part provides general information for outside branch circuits and feeders, including the following:
- Conductor Size and Support
- Outdoor Lighting Equipment
- Overcurrent Protection
- Wiring on Buildings
- Entrance/Exit of Circuits
- Open Conductor Spacing/Support
- Support Over Buildings
- Point of Attachment
- Means of Attachment
- Clearance from Ground
- Protection of Conductors
- Raceways/Cables on Exterior
- Vegetation as Support

II. Buildings or Other Structures Supplied by a Feeder(s) or Branch Circuit(s)
This part addresses outside branch circuits and feeders which supply a separate building or structure. Note that a building or structure supplied by an outdoor branch circuit or feeder must comply

with rules very similar to those for a service-supplied building or structure.

This part also provides requirements for outside branch circuits and feeders which supply more than one building or structure. These basic requirements include the following:

- Number of Supplies
- Location of Disconnecting Means
- Maximum Number of Disconnects
- Grouping of Disconnects
- Accessibility of Disconnects
- Identification of Disconnects
- Disconnect Rating
- Access to Overcurrent Protective Devices

III. Over 600 Volts

Outside branch circuits and feeders operating at over 600 volts are addressed in **Part III** of **Article 225**. This part provides requirements very similar to those in **Part I** and **Part II**, but modified for systems operating at over 600 volts.

Article 230 Services

Proper application of **Article 230** requires that the *Code* user be familiar with several **Article 100** definitions related to services. **See Figure 8-8.**

> **Service.** The conductors and equipment for delivering electric energy from the serving utility to the wiring system of the premises served.

This definition encompasses all conductors from the service point to the service disconnect, and all overcurrent protection. This scope is also seen in the definition of service conductors.

> **Service Conductors.** The conductors from the service point to the service disconnecting means.

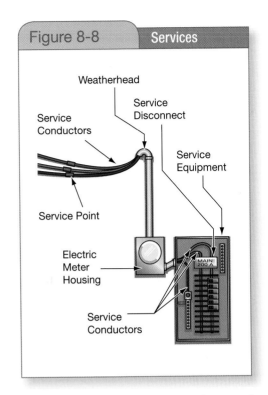

Figure 8-8	Services

Figure 8-8. Article 230 covers all conductors and equipment from the service point to the service disconnect and overcurrent protection.

This definition would include service drop (overhead) and service lateral (underground) conductors. The requirements of **Article 230** have been specifically developed to protect persons and property from service conductors which are considered to be unprotected and, in most cases, can only be disconnected by the serving utility.

Article 230 is divided into eight logical parts, with each part logically named for the requirements contained within:

I. General
- Number of Services
- Service Conductors Considered as Outside
- Raceway Seals
- Clearance of Service Conductors from Building Openings
- Vegetation as Support

II. Overhead Service Conductors

- Insulation or Covering
- Conductor Size and Rating
- Clearance Above Ground and Rooftops
- Point of Attachment to Building
- Means of Attachment
- Service Masts as Supports
- Supports over a Building

III. Underground Service Conductors

- Insulation
- Conductor Size and Rating
- Splices
- Protection

IV. Service-Entrance Conductors

- Number of Sets Served
- Insulation
- Protection
- Conductor Size and Rating
- Permitted Wiring Methods
- Splices
- Cable Trays
- Supports
- Raceways Arranged to Drain
- Overhead Connections Service Head/ Gooseneck, Drip-loops
- 4-wire Delta Configured Systems

V. Service Equipment – General

- Enclosure Requirements
- Marking of Equipment, Suitable as "Service Equipment"

VI. Service Equipment – Disconnecting Means

- Disconnect Requirement
- Location of Disconnect Marking
- Accessibility of Disconnect
- Maximum Number of Disconnects
- Grouping of Disconnects
- Operation of Disconnects
- Indicating (Identifies Open or Closed)

- Rating of Disconnect (amps)
- Equipment Permitted on the Line Side (Upstream) of the Service Disconnect

VII. Service Equipment - Overcurrent Protection

- Where Required
- Location
- Locking of Service OCPD
- Ground-Fault Protection of Equipment (GFPE)

VIII. Services Exceeding 600 Volts, Nominal

- General Requirements
- Isolating Switches
- Disconnects
- Protection
- Surge Arresters
- Metal Enclosed Switchgear
- Services over 35,000 Volts

PROTECTION

In the planning stages of any electrical installation, protection must be incorporated in the electrical drawings as part of the design. The drawings provide the installer with the necessary information to build an installation which is properly protected. All electrical installations must provide overcurrent protection for all current-carrying conductors, which must be protected in accordance with the conductor ampacity and conditions of use in accordance with **Article 240 Overcurrent Protection**.

Grounding of electrical systems is designed to limit the voltage imposed by lightning, line surges, or unintentional contact with higher voltage lines and to stabilize the voltage to earth during normal operation. Normally non–current-carrying metal parts of electrical systems are connected to together (bonded) and are connected to the electrical supply source in a manner that creates an effective ground-fault current path. Systems and circuits are required to be grounded in accordance with **Article 250 Grounding and Bonding**. Protection from surge

> **Fact**
>
> An exception has been added to 230.24(A) of the 2011 *NEC*. The exception permits clearance between a roof and service conductors operating at not more than 300 volts to not less than 3 feet where the roof area is guarded or isolated.

voltages many times attributed to lightning is provided in electrical systems through the use of **Article 280 Surge Arrester, Over 1 kV**. Protection from voltage surges at levels closer to nominal voltage is provided for by the application of **Article 285 Surge-Protective Devices (SPDs), 1 kV or Less**.

Code-compliant electrical installations must provide adequate protection of persons and property in accordance with the *NEC*. The required protection for electrical installations which must be planned include the following keywords/clues:

- Overcurrent protection, fuses, circuit breakers
- Grounding and bonding of systems and circuits
- Grounding and bonding raceways, equipment, or enclosures
- Surge arresters, over 1 kV
- Surge-protective devices (SPDs), 1 kV or less

Articles dedicated to protection can also be found in Chapter 2. **See Figure 8-9**.

Article 240 Overcurrent Protection

Article 100 defines the term *overcurrent*.

> **Overcurrent.** Any current in excess of the rated current of equipment or the ampacity of a conductor. It may result from overload, short circuit, or ground fault.

This basic definition must be understood to properly apply the provisions of **Article 240**. All conductors and equipment are rated for the maximum amount of current (amps) that it can handle without suffering damage. An overcurrent occurs when one of three incidents happens: an overload, a short circuit, or a ground fault.

Overloads occur when a conductor or equipment is subjected to a current exceeding its ampere rating. Overload cur-

Figure 8-9	Chapter 2 (Wiring and) Protection
NEC® Title: Codeology Title: Chapter Scope:	Wiring and Protection Plan Information and Rules on *Wiring and Protection* of Electrical Installations

PROTECTION	
Article	**Article Title**
240	Overcurrent Protection
250	Grounding and Bonding
280	Surge Arresters, Over 1 kV
285	Surge-Protective Devices (SPDs), 1 kV or Less

Figure 8-9. Articles 240, 250, 280, 285 cover the protection of Chapter 2.

For additional information, visit qr.njatcdb.org Item #1066

rent stays on the normal circuit path which, if allowed to continue for too long, would cause damage or dangerous overheating. A fault, such as a short circuit or ground fault, is not an overload. A conductor rated at 20 amps with 22 amps of current flowing would be experiencing an overload. An overload never leaves the circuit path. Current continues to flow from the source through the circuit conductors and load back to the source.

A short circuit occurs when the current leaves the normal circuit path and takes a shortcut back to the source. Short circuits occur when current-carrying conductors make contact with each other, thus creating a shortcut for current flow. Any combination of two or more circuit conductors (current-carrying) in contact results in a short circuit. Current-carrying conductors include all ungrounded (hot) and grounded (in most cases, neutral) conductors.

A ground fault is a form of short circuit. The current takes a shortcut on a grounded component, such as a raceway, enclosure, building steel, or an equipment grounding or bonding conductor.

The scope of **Article 240 Overcurrent Protection** is to provide general requirements for overcurrent protection and for

Figure 8-10   OCPDs

Figure 8-10. Fuses and circuit breakers provide overcurrent protection for conductors and equipment.

For additional
information, visit
qr.njatcdb.org
Item #1067

overcurrent protective devices. Overcurrent protective devices (OCPDs) include, but are not limited to, fuses and circuit breakers. **See Figure 8-10**. These devices are designed to provide protection from overloads, short circuits, and ground faults.

Article 240 consists of nine logical parts as follows:

I. General
- Definitions
- Cross-Reference Other Articles
- Protection of Conductors
- Protection of Flexible Cords/Cables
- Standard OCPD amp Ratings
- Fuses or Circuit Breakers in Parallel
- Supplementary Overcurrent Protection
- Thermal Devices
- Electrical System Coordination
- Ground-Fault Protection of Equipment

II. Location
- Ungrounded Conductors
- Location of Overcurrent Protection in Circuit, Tap Rules
- Grounded Conductors
- Location/Accessibility of OCPDs

III. Enclosures
- General Protection and Operation
- Damp or Wet Locations
- Vertical Position

IV. Disconnecting and Guarding
- Disconnects for Fuses
- Arcing or Suddenly Moving Parts

V. Plug Fuses, Fuseholders, and Adapters
- General Application
- Edison Base Fuses
- Type "S" Fuses

VI. Cartridge Fuses and Fuseholders
- General Application
- Classification

VII. Circuit Breakers
- Method of Operation
- Indicating
- Nontamperable
- Marking
- Applications
- Series Ratings

VIII. Supervised Industrial Installations
This part addresses only those portions of a building or structure that meet the conditions of a supervised industrial installation, as defined in **240.2**.

IX. Overcurrent Protection Over 600 Volts, Nominal
This part is limited only to feeders and branch circuits operating at over 600 volts nominal.

Article 250 Grounding and Bonding
The scope of **Article 250** is provided in the first section of the article. **See Figure 8-11.**

Figure 8-11   Grounding and Bonding

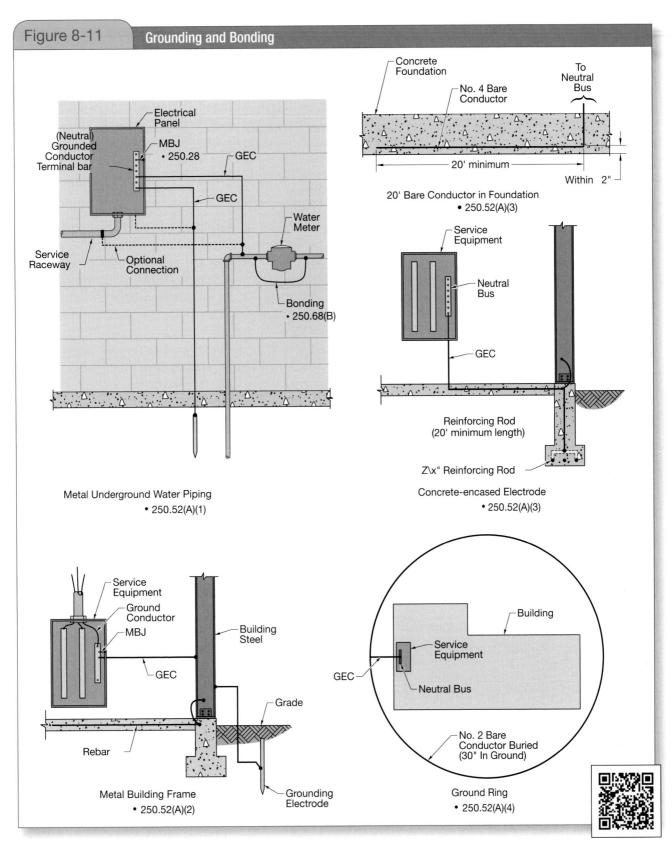

Figure 8-11. *Article 250 covers protection requirements through provisions for grounding and bonding of electrical systems and installations.*

For additional information, visit qr.njatcdb.org Item #1068

250.1 Scope. This article covers general requirements for grounding and bonding of electrical installations as well as the specific requirements found in (1) through (6).

(1) Systems, circuits, and equipment required, permitted, or not permitted to be grounded

(2) Circuit conductor to be grounded on grounded systems

(3) Location of grounding connections

(4) Types and sizes of grounding and bonding conductors and electrodes

(5) Methods of grounding and bonding

(6) Conditions under which guards, isolation, or insulation may be substituted for grounding

The *NEC* is a "prescriptive" installation document, meaning that it makes specific compliance requirements of the installer without stating the reason(s) for that compliance. The *NEC* does not explain individual requirements for two reasons:

1. Per **90.1(C)**, the *NEC* is not intended to be a design or instruction manual for untrained persons
2. The volume of the *NEC* precludes such explanations

If all requirements were explained, the *NEC* would be five times thicker in size. However, in **Article 250 Grounding and Bonding, 250.4** explains to the user exactly what grounding and bonding of electrical systems are required to accomplish.

250.4 General Requirements for Grounding and Bonding

(A) Grounded Systems

(1) Electrical System Grounding

(2) Grounding of Electrical Equipment

(3) Bonding of Electrical Equipment

(4) Bonding of Electrically Conductive Materials and Other Equipment

(5) Effective Ground-Fault Current Path

(B) Ungrounded Systems

(1) Grounding Electrical Equipment

(2) Bonding of Electrical Equipment

(3) Bonding of Electrically Conductive Materials and Other Equipment

(4) Path for Fault Current

Article 250 is subdivided into ten logical parts as follows:

I. General
- Definitions
- Cross-Reference Other Articles
- Reasons for Grounding and Bonding Connections
- Protection of Clamps/Fittings
- Clean Surfaces

II. System Grounding
- Systems Requiring Grounding
- Systems NOT Requiring Grounding
- Circuits not Permitted to Be Grounded
- Grounding AC Services
- Conductor to be Grounded
- Main Bonding Jumpers
- Separately Derived Systems
- Two or More Buildings with Common Service
- Portable Generators
- High-Impedance Systems

III. Grounding Electrode System and Grounding Electrode Conductor
- Outline of the Grounding Electrode System
- Permitted Electrodes
- Installation of Grounding Electrode System
- Common Electrodes
- Supplementary Electrodes
- Resistance of Electrodes
- Air Terminals
- Size of Grounding Electrode Conductor
- Connection of Grounding Electrode Conductors

IV. Enclosure, Raceway, and Service Cable Connections
- Service Raceways and Enclosures
- Underground Service Cable and Conduit
- Other Enclosures and Raceways

V. Bonding
- Services
- Other Systems
- Other Enclosures
- Over 250 Volts
- Loosely Jointed Raceways
- Hazardous Locations
- Equipment Bonding Jumpers, Supply and Load Side
- Piping Systems and Exposed Structural Steel
- Lightning Protection Systems

VI. Equipment Grounding and Equipment Grounding Conductors
- Equipment Fastened in Place
- Cord-and-Plug-Connected Equipment
- Nonelectric Equipment
- Types of Equipment Grounding Conductors
- Identification of Equipment Grounding Conductors and Device Terminals
- Installation
- Size of Equipment Grounding Conductors

VII. Methods of Equipment Grounding
- Connections
- Short Sections of Raceway
- Ranges/Clothes Dryer Frames
- Equipment Fastened in Place
- Cord-and-Plug-Connected Equipment
- Use of Grounded Conductor
- Receptacle Grounding Attachment to Boxes

VIII. Direct-Current Systems
- Circuits and Systems to Be Grounded
- Point of Connection
- Size of Grounding Electrode Conductor
- Bonding Jumpers

IX. Instruments, Meters, and Relays
- Transformer Circuits and Cases
- Cases of Equipment at Over 1,000 Volts
- Grounding Conductors

X. Grounding of Systems and Circuits over 1 kV
- General Requirements
- Derived Neutral Systems
- Grounding
- Solidly Grounded Neutral Systems
- Impedance Grounded Neutral Systems
- Portable or Mobile Equipment

Article 280 Surge Arresters, Over 1 kV
Surge arrester is defined in **Article 100**.
See Figure 8-12.

Surge Arrester. A protective device for limiting surge voltages by discharging or bypassing surge current; it also prevents continued flow of follow current while remaining capable of repeating these functions.

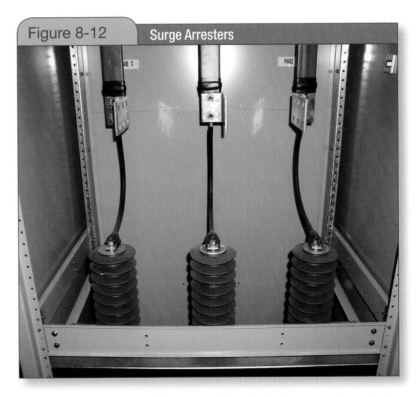

Figure 8-12   Surge Arresters

Figure 8-12. Surge arresters limit surge voltages by discharging or bypassing the dangerous surge voltages to ground.

Article 280 consists of three logical parts as follows:

I. General
- Uses Not Permitted
- Listing
- Number of Surge Arresters Required
- Selection

II. Installation
- Location of Surge Arresters
- Routing Surge Arrester Connections

III. Connecting Surge Arresters
- Services of Over 1,000 Volts
- Load Side Installation of Over 1,000 Volts
- Circuits 1,000 Volts and Over
- Grounding

Article 285 Surge-Protective Devices (SPDs), 1 kV or Less

Article 100 defines a *surge-protective device* (SPD). **See Figure 8-13.**

Figure 8-13 | SPDs 1 kV or Less

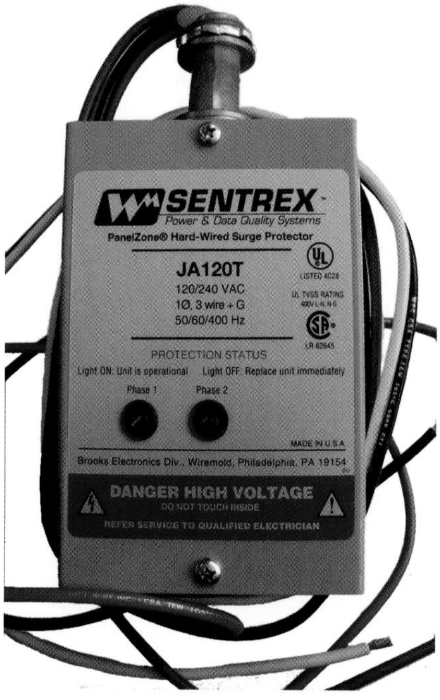

Figure 8-13. Surge-protective devices (SPDs) provide protection from an overvoltage at levels much closer to the operating voltage that surge arresters.

Surge-Protective Device (SPD). A protective device for limiting transient voltages by diverting or limiting surge current; it also prevents continued flow of follow current while remaining capable of repeating these functions and is designated as follows:

Type 1: Permanently connected SPDs intended for installation between the secondary of the service transformer and the line side of the service disconnect overcurrent device.

Type 2: Permanently connected SPDs intended for installation on the load side of the service disconnect overcurrent device, including SPDs located at the branch panel.

Type 3: Point of utilization SPDs.

Type 4: Component SPDs, including discreet components, as well as assemblies.

Article 285 consists of three logical parts as follows:

I. General
- Listing
- Uses Not Permitted
- Number Required
- Listing Requirements
- Short Circuit Ratings

II. Installation
- Location
- Routing of Connections

III. Connecting SPDs
- Connection of SPDs
- Grounding

Fact

If an SPD is provided with a cautionary marking indicating specific installation requirements, installers are required to comply with all those installation instructions included in the listing and labeling of electrical equipment.

KEY WORDS AND CLUES FOR CHAPTER 2, "PLAN"

A list of key words and clues for Chapter 2 are included for reference. **See Figure 8-14.** Chapter 2 topics are many of the most common *Code* requirements in daily installations. As previously discussed, grounding and bonding applies to practically every electrical installation. It is paramount that electricians have a solid understanding of Chapter 2 and how to locate sections based upon key words and clues.

Refer to Chapter 2 when the question or need within the *NEC* deals with any of the following:

- Wiring, *NEC* current-carrying conductor names, branch circuit, feeder, service or tap conductor
- Grounded Conductors
- Ground-Fault Circuit Interrupters (GFCIs)
- Arc-Fault Circuit Interrupters (AFCIs)
- Required Outlets
- Calculations, Computed Loads
- Outside Branch Circuits and Feeders
- Protection, Overcurrent, Grounding, Surge Arresters, and Surge-Protective Devices
- Fuses and Circuit Breakers
- Grounding and Bonding Conductors
- Grounding Electrodes

Figure 8-14	Key Words and Clues
WIRING	**PROTECTION**
Branch circuits, all types GFCI	Overcurrent protection, fuses, circuit breakers
Branch circuits, indoor/outdoor	Grounding systems and circuits
Required outlets AFCI	Grounding and bonding raceways, equipment or enclosures
Feeders, indoor/outdoor	Surge arresters
Service/s service equipment	Surge-protective devices
Calculations, computed load	Equipment grounding conductors
Grounded conductors, neutral	Location of overcurrent protective devices

Figure 8-14. *Key words and clues regarding wiring and protection will lead the Code user to review Chapter 2 of the NEC.*

Summary

Chapter 2, in accordance with **Section 90.3**, applies generally to all electrical installations. Using the *Codeology* method, this chapter has been coined the "Plan" chapter, due to its wide scope of coverage. The title of **Chapter 2 Wiring and Protection** has been described as information and rules on wiring and protection of electrical installations.

Chapter 2 covers the entire electrical system from the service point (connection to the utility) to the last receptacle or other outlet in the electrical system. Chapter 2, in accordance with its scope of "Wiring," provides detailed requirements for all current-carrying conductors. To provide clarity, the *NEC* names and defines these conductors. Names include service, feeder, branch-circuit, and tap conductors.

In accordance with its scope of "Protection," Chapter 2 provides detailed requirements to protect the entire electrical system. These protection requirements include overcurrent, grounding, bonding, surge arresters, and surge-protective devices.

Chapter 2, along with Chapters 1, 3, and 4, builds the foundation or backbone of all electrical installations. **See Figure 8-15.**

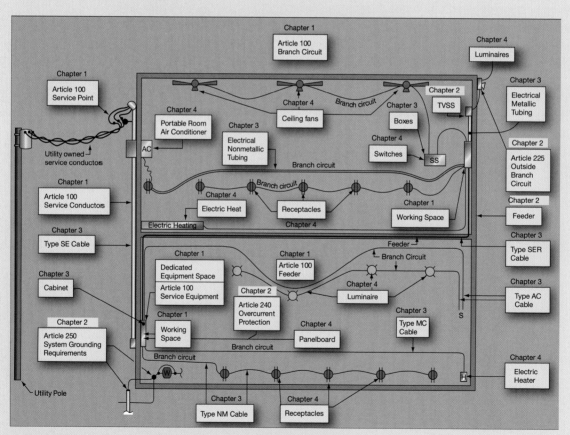

Figure 8-15. Chapter 2 will apply generally in all electrical installations.

Review Questions

1. Does Chapter 2 of the *NEC* apply to all electrical installations covered by the *Code*?

2. Chapter 2 is subdivided into how many articles?

3. The scope of Chapter 2 of the *NEC* consists of six articles covering __?__ and four articles covering __?__ .

4. Name the part and article of Chapter 2 which addresses an outdoor branch circuit serving a second structure.

5. Chapter 2 of the *NEC* addresses wiring, which would include calculating the size of conductors. Which part of which article would apply when calculating loads for a single-family dwelling unit service?

6. Does 210.8(B) address ground fault circuit interrupter requirements in dwelling units?

7. The location of 250.166 limits the application of this section to what type of system?

8. Article 240 is divided into how many parts?

9. Does Part V of Article 240 apply to cartridge-type fuses?

10. Which part of Article 250 would contain requirements for grounding electrode conductors?

To access Practice Problems, visit qr.njatcdb.org. Item #1082 Click on Inside Blended Learning.

Chapter 3 of the *NEC*, "BUILD"

The *Codeology* title for Chapter 3 is "Build." Electrical installations can only occur after they have been properly planned. Chapter 3 provides the rules and information necessary for all of the wiring methods and materials used to distribute electrical energy. Electrical equipment, such as switches, receptacles, panelboards, switchboards, motors, lighting, and appliances, are not contained within the scope of this chapter. These types of equipment, which provide for control, transformation, and utilization, are addressed in other chapters of the *NEC*.

Objectives

» Associate the *Codeology* title for *NEC* Chapter 3 as "Build."

» Understand the building and hands-on type of information and requirements for wiring methods and wiring materials contained in Chapter 3.

» Identify Chapter 3 numbering as the 300-series.

» Recognize, recall, and apply the articles contained in Chapter 3.

Chapter 9

Table of Contents

NEC CHAPTER 3, "BUILD"

Chapter 3, the 300-series, contains forty-five articles. These articles address the methods and materials for the distribution of electrical energy. In essence, Chapter 3 covers all methods and materials necessary to connect electrical energy from the power source to the electrical equipment. This chapter could be considered the mechanical work an electrician installs, or the mechanical distribution means by which electrical energy is delivered; often referred to as "pipe and wire." The wiring methods and materials covered in the "Build" chapter include the following:

• Conductors
• Enclosures, cabinets, boxes
• Cable assemblies
• Circular raceways, conduits
• Other raceways
• Busways/cablebus
• Open wiring

The materials used to distribute electrical energy throughout buildings and structures are broken down to *wiring methods* and *wiring materials*. These two breakdowns are basically the root of Chapter 3. *Wiring methods* are the materials which carry the current (i.e. conductors) along the mechanical means to carry and route these conductors (i.e. conduits). *Wiring materials* facilitate the electrical installation by utilizing items such as junction boxes, hangers, and other support items. Chapter 3 of the *NEC* covers various types of conductors and cables, raceways (i.e. conduit), and junction boxes which are used in electrical installations. **See Figure 9-1.**

The requirements and information in *NEC* Chapter 3 are logically separated

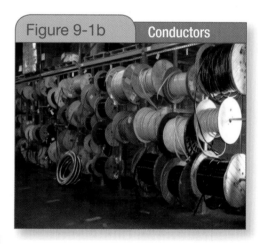

Figure 9-1b. *Conductors (Wiring Methods) of various sizes and insulation levels are often installed together.*

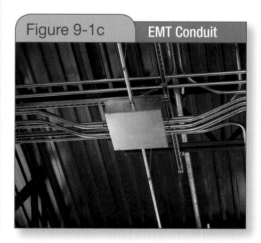

Figure 9-1c. *EMT conduit (Wiring Methods) is often used to distribute power in commercial installations.*

Figure 9-1a. *Junction boxes (Wiring Materials), as listed in Article 314, are commonly used in the distribution of the conduit system which provides electrical service to outlets and equipment.*

into 13 different categories. **See Figure 9-2.** Practically every electrical installation requires current-carrying conductors from the power source to the final equipment. These conductors must be protected and routed in a physically protective means, typically by a conduit or cable covering.

WIRING METHODS

Wiring methods are actually the materials used to conduct current throughout an electrical installation. Any combination of conductors and a protective means or layer to facilitate installation is a wiring method. Very often the conductors are grouped together under one protective means or layer which are referred to as a cable assembly (or simply called a cable). For further clarification, a conductor is a single wire, whereas a cable is composed of multiple wires under one protective means or layer.

Cable assemblies are common wiring methods. **See Figure 9-3.** For example, type AC cable consists of insulated conductors, wrapped in paper, with an outer

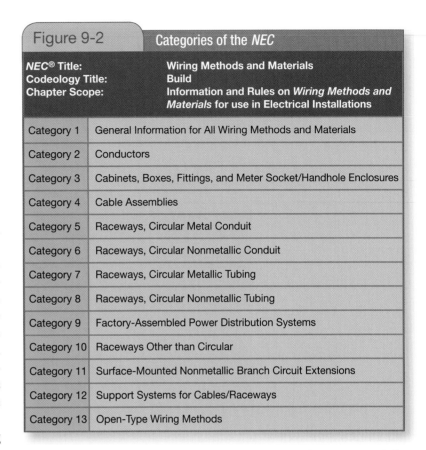

Figure 9-2	Categories of the *NEC*

NEC® Title:	**Wiring Methods and Materials**
Codeology Title:	**Build**
Chapter Scope:	**Information and Rules on *Wiring Methods and Materials* for use in Electrical Installations**

Category 1	General Information for All Wiring Methods and Materials
Category 2	Conductors
Category 3	Cabinets, Boxes, Fittings, and Meter Socket/Handhole Enclosures
Category 4	Cable Assemblies
Category 5	Raceways, Circular Metal Conduit
Category 6	Raceways, Circular Nonmetallic Conduit
Category 7	Raceways, Circular Metallic Tubing
Category 8	Raceways, Circular Nonmetallic Tubing
Category 9	Factory-Assembled Power Distribution Systems
Category 10	Raceways Other than Circular
Category 11	Surface-Mounted Nonmetallic Branch Circuit Extensions
Category 12	Support Systems for Cables/Raceways
Category 13	Open-Type Wiring Methods

Figure 9-2. Articles of Chapter 3 describe wiring methods and wiring materials for electrical installations.

Figure 9-3a	Wiring Methods

Figure 9-3a. Wiring methods using nonmetallic-sheathed cables (NM) are typically used in residential installations.

armor of flexible corrugated metal outer sheath of aluminum or steel that includes an internal bonding strip of copper or aluminum in intimate contact with the armor for its entire length. Eleven types of cable assemblies are covered in Chapter 3.

Raceways with conductors installed are also wiring methods. For example, electrical metallic tubing, type EMT with insulated type THHN conductors installed, is a wiring method. Chapter 3 covers 25 types of raceways in five different categories. **See Figure 9-4.**

The decision of which of the five categories of raceway to use in an installation is based upon several factors such as economics, required physical protections, environmental location, and even AHJ requirements. In some cases, it may be based upon what the customer desires, as long as it permissible under the *NEC*. The installation practices are certainly different between metal and PVC raceways. More than one type of raceway is often used on an electrical project. **See Figure 9-5.**

Figure 9-3b | **Wiring Methods**

Figure 9-3b. Wiring methods using metal-clad cables (MC) are typically used in commercial installations.

Figure 9-5a | **RMC (Article 344)**

Figure 9-5a. Circular metal conduit provides maximum physical protection.

Figure 9-4 | **Raceway Categories**

Categories of Raceways	Figure Reference
Raceways, Circular Metal Conduit	Figure 9-5a
Raceways, Circular Nonmetallic Conduit	Figure 9-5b
Raceways, Circular Metallic Tubing	Figure 9-5c
Raceways, Circular Nonmetallic Tubing	Figure 9-5d
Raceways Other than Circular	Figure 9-5e

Figure 9-4. Various means of raceways are utilized in electrical installations.

Figure 9-5b | **PVC (Article 352)**

Figure 9-5b. Circular nonmetallic conduit is mainly utilized underground and outside.

Figure 9-5c. Circular metal tubing is popular in commercial and light industrial areas.

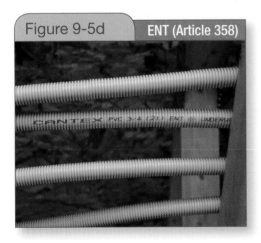

Figure 9-5d. Circular nonmetallic tubing has become increasing popular in commercial installations.

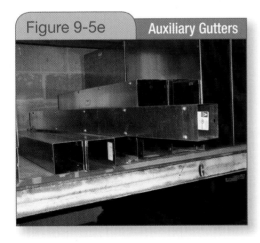

Figure 9-5e. Non-circular raceways are especially used in industrial areas where a high quantity of conductors are employed.

For example, a branch circuit run feeding an electric sign in the front yard of an elementary school may start from an indoor electrical panel located in the school office. As the conduit run is located above the office ceiling, EMT (**Article 358**) conduit is commonly permissible. As the conduit penetrates the outside wall it may be routed under the bus stop canapé for a distance where it may be subject to physical abuse. For this portion of the raceway run, RMC (**Article 344**) is likely required. After the conduit runs down the canapé post and continues underground out to the sign, PVC (**Article 352**) is commonly utilized.

Busways and cablebus are factory-assembled sections of grounded, completely enclosed, ventilated protective metal housings containing factory-mounted, bare or insulated conductors, which are usually copper or aluminum bars, rods, or tubes. **See Figure 9-6.**

Open wiring methods consist of concealed knob and tube wiring and open wiring on insulators. Note that these open wiring methods are without a protective outer jacket or enclosure and are extremely limited in application. This practice is not utilized in new installation but can be found in older maintenance situations.

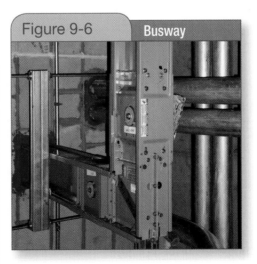

Figure 9-6. Busways are used when a large current capacity is required and there is a need to supply multiple pieces of equipment within close proximity.

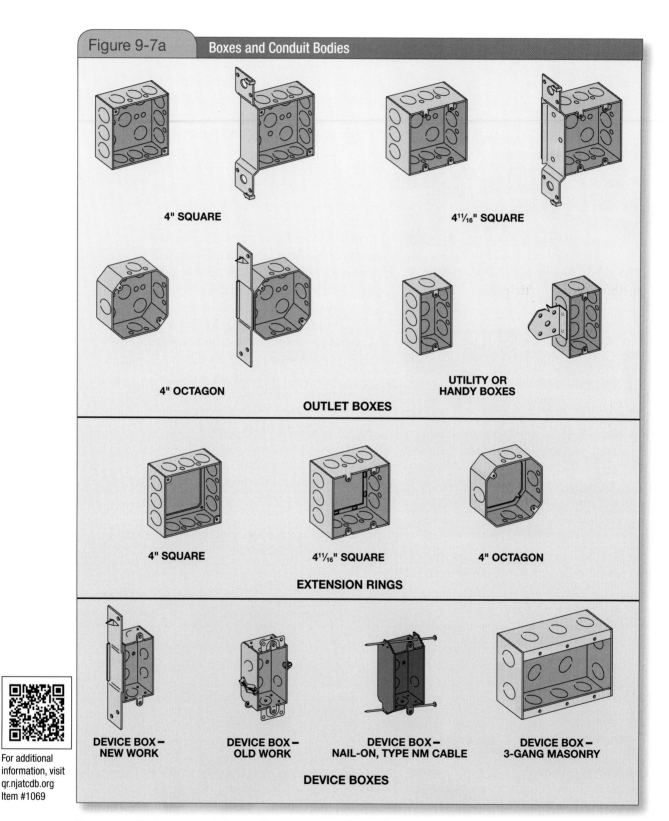

Figure 9-7a | Boxes and Conduit Bodies

4" SQUARE

4¹¹⁄₁₆" SQUARE

4" OCTAGON

UTILITY OR HANDY BOXES

OUTLET BOXES

4" SQUARE

4¹¹⁄₁₆" SQUARE

4" OCTAGON

EXTENSION RINGS

DEVICE BOX – NEW WORK

DEVICE BOX – OLD WORK

DEVICE BOX – NAIL-ON, TYPE NM CABLE

DEVICE BOX – 3-GANG MASONRY

DEVICE BOXES

For additional information, visit qr.njatcdb.org Item #1069

Figure 9-7a. Various sizes and shapes of boxes are required in an electrical installation.

WIRING MATERIALS

Wiring materials facilitate the installation of wiring methods. For example, enclosures and boxes are necessary for the installation of all types of wiring methods and equipment. Chapter 3 covers cabinets, cutout boxes, meter socket enclosures, outlet boxes, device boxes, pull boxes, junction boxes, conduit bodies, fittings, and handhole enclosures. **See Figure 9-7.**

Support systems are a type of wiring material. Messenger-supported wiring is a means of support for a given wiring method listed in **Article 396**. Cable trays are considered Wiring Materials and are not considered a Wiring Method. They are not listed as raceways in the *NEC*. Cable Trays are utilized to support raceways, cables, and single conductors as covered in **Article 392**. **See Figure 9-8.**

GENERAL INFORMATION FOR ALL WIRING METHODS AND MATERIALS

Article 300 Wiring Methods

The first article in Chapter 3, **Article 300 Wiring Methods** covers the general rules for all wiring methods and wiring materials. All electrical installations must conform to these general requirements, unless modified by another article in Chapter 5, 6, or 7. Thus, **Article 300** is the backbone of the "Build" chapter. General requirements for all wiring methods and materials for each raceway or cable assembly article are listed in **Article 300** rather than being repeated in each individual section.

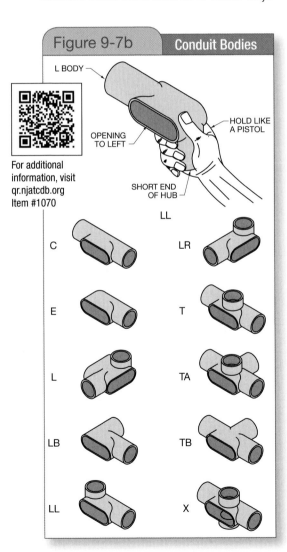

Figure 9-7b. Conduit bodies are required by the NEC to facilitate conductor installations.

Figure 9-8. Cable trays come in various sizes and styles. This particular cable tray is often referred to as a ladder bottom tray. Many times the cable tray will have a solid bottom.

The key provisions of **Article 300 Wiring Methods** include the following sections:

300.4 Protection Against Physical Damage requires that all conductors, raceways, and cables subject to physical damage be protected. Specific requirements are given for wiring methods through framing members and for ungrounded (hot) conductors 4 AWG (American Wire Gage) or larger entering a cabinet, a box, an enclosure, or a raceway.

300.5 Underground Installations and **Table 300.5 Minimum Cover Requirements, 0 to 600 Volts, Nominal, Burial in Millimeters (Inches)** contain specific requirements for all underground installations of direct-buried cables and raceways.

300.11 Securing and Supporting requires that all raceways, cable assemblies, boxes, cabinets, and fittings be securely fastened in place.

Additional requirements cover wiring located within floor, ceiling, and roof assemblies. Specific support requirements for raceways and for cable assemblies for support spacing are included in the raceway or cable assembly article. Keep in mind, many local electrical licensing ordinances will require various levels of seismic anchoring of raceways and equipment.

300.14 Length of Free Conductors at Outlets, Junctions, and Switch Points requires a minimum length of conductor from boxes and enclosures to allow for splicing or device termination. The only exception to this requirement is if conductors are not spliced or terminated at the outlet, junction, or switch point.

300.15 Boxes, Conduit Bodies, or Fittings—Where Required requires that, in general, when raceways or cable assemblies are used, a box or conduit body (an LB, for example) must be installed at each conductor splice point and at each outlet, switch, junction, and termination point.

300.21 Spread of Fire or Products of Combustion requires that when wiring methods or materials penetrate an opening through fire-resistant walls, partitions, floors, or ceilings, the opening be installed with fire-stopped materials in accordance with approved methods to maintain their fire resistance rating.

300.22 Wiring in Ducts Not Used for Air Handling, Fabricated Ducts for Environmental Air, and Other Spaces for Environmental Air (Plenums) specifically addresses permitted wiring methods and materials in ducts, plenums, and other air-handling spaces.

Code users must become extremely familiar with the general requirements of **Article 300**, given their required application to every electrical installation regardless of the materials used. **Article 300** is divided into two logical parts as follows:

I. General Requirements

- Voltage and Temperature Limitations
- Single Conductors
- Grouping of Conductors of the Same Circuit
- Paralleled Conductors of Different Systems
- Protection of Conductors
- Underground Installations
- Protection from Corrosion and Deterioration
- Raceways Exposed to Different Temperatures
- Electrical Continuity of Raceways
- Securing and Supporting
- Mechanical and Electrical Continuity of Conductors
- Where Boxes, Conduit Bodies, and Fittings Are Required
- Transition from Raceway or Cable to Open or Concealed Wiring
- Number and Size of Conductors in a Raceway
- Raceway Installations, General
- Supporting Conductors in Vertical Raceways

- Induced Currents in Metal Enclosures or Raceways
- Spread of Fire or Products of Combustion
- Wiring in Ducts, Plenums, and Other Air-Handling Spaces
- Panels Designed to Allow Access

II. Requirements for over 600 Volts, Nominal

- Required Covers
- Conductors of Different Systems
- Conductor Bending Radius
- Protection Against Induction Heating
- Aboveground Wiring Methods
- Braid-Covered Insulated Conductors
- Insulation Shielding
- Moisture and Mechanical Protection for Metal-Sheathed Cables
- Underground Installations

CONDUCTORS

Article 310 Conductors for General Wiring

All general requirements for conductors in cable assemblies, raceways, open wiring, or support systems are contained in **Article 310 Conductors for General Wiring.** Over the past few *Code* cycles, many of the sections within **Article 310** have been relocated and revised. If you were somewhat familiar with **Article 310**, be sure to study it again for these relocations and changes. Also keep in mind that specific cables such as flexible cords are covered in **Article 400**, and specific fixture wires in **Article 402**.

The key provisions of **Article 310 Conductors for General Wiring** cover the following:

1. Type designations, such as THWN or XHHW, are part of the required marking for all conductors. These letters and suffixes, used in the type designation to explain the physical properties of the conductor, are extremely important to the *Code* user. The *NEC*

discusses the labeling requirement to include: trade name, maximum operating temperatures, applications, insulation, thickness of insulation, and outer covering for the individual types of conductors. **See Figure 9-9.**

Conductor Type Designations	
T	Thermoplastic insulation
R	Thermoset insulation
S	Silicone (thermoset) insulation
X	Cross-linked synthetic polymer insulation
Z	Modified tetrafluoroethylene insulation
U	Underground use
L	Lead sheath
N	Nylon jacket
W	Moisture resistant
H	75°C rated (lack of "H" usually indicates 60°C rating)
HH	90°C rated
-2	The suffix "-2" designates continuous 90°C rating, wet or dry

2. Insulation requirements for all conductors specify the type and thickness of all conductor insulation. Types of conductor insulation include, but are not limited to, TW, THW, THHW,

Figure 9-9   Conductor Markings

Figure 9-9. Article 310 covers general requirements for conductors and their type designations, insulations, markings, mechanical strengths, ampacity ratings, and uses.

THHN, RHH, RHW, and XHHW. The marking requirements for conductors ensure that all conductors and cables are marked to indicate the maximum rated voltage, the proper type designation, the manufacturer's name/trademark, and the size of the conductor in accordance with the circular mil area (cmil) or the American Wire Gage (AWG).

3. Mechanical strengths of conductors are addressed by requirements for minimum size, thickness of insulation, outer coverings, and permitted applications.

4. Ampacity ratings for all types and uses of conductors are included in the ampacity tables in **Article 310**. Each of these tables addresses different installation possibilities for all types of conductors in all ranges of temperature limitations and installation methods. **Table 310.15(B)(16)**, however, is the most frequently used ampacity table. **310.15** provides specific requirements for conductor ampacity and the corrections for exceeding the number of current-carrying conductors permitted in the tables. Type designation letters, such as type THWN, are used to determine the ampacity of the conductor using the proper table.

Uses for all type designations of conductors are detailed in **Table 310.104(A)** and **Table 310.104(B)**. For example, type THHW is permitted for use in dry locations with a temperature limitation of 90°C, and in wet locations with a temperature limitation of 75°C.

Key provisions of **Article 310 Conductors for General Wiring** include the following:

310.10(A) Dry Locations permits all conductors to be installed in dry locations.

310.10(B) Dry and Damp Locations provides a list of type designations permitted in damp and wet locations.

310.10(H) Conductors in Parallel requires that, in general, only conductors 1/0 AWG or larger be permitted to be installed in parallel.

310.15 Ampacities for Conductors Rated 0–2,000 Volts provides general information on conductor ampacity and specific requirements for ampacity adjustment.

310.15(A)(3) Temperature Limitation of Conductors requires that the temperature limitation determined by the type designation for all conductors not be exceeded.

310.106(C) Stranded Conductors requires all conductors 8 AWG and larger and installed in raceways to be stranded.

310.120 Marking details required marking information and the methods permitted for marking.

Code users must become familiar with the general requirements listed in **Article 310**, given their broad application to every electrical installation, regardless of the conductors used. **Article 310** is subdivided into parts with requirements for the following:

I. General
- Scope
- Definitions

II. Installation
- Conductors in dry, damp, and wet locations
- Direct-Burial Conductors
- Corrosive Conditions
- Conductors in Parallel
- Equipment Bonding Jumpers
- Ampacities for Conductors
- Temperature Limitation of Conductors
- Bare and Neutral Conductors
- Conductors Rated 2,001–35,000 Volts

III. Construction Specifications
- Conductor Construction and Application
- Stranded Conductors
- Conductor Identification
- Minimum Size of Conductors
- Marking

Cabinets, Boxes, Fittings, and Meter Socket/Handhole Enclosures

To facilitate the installation of all types of wiring methods, the need arises for wiring materials. Cabinets, boxes, fittings, and meter sockets, handholes, and enclosures are the wiring materials which are common to many electrical installations. The two articles in Chapter 3 which cover these wiring materials are **Article 312** and **Article 314**.

Article 312 Cabinets, Cutout Boxes, and Meter Socket Enclosures

Article 312 provides general rules for the installation and construction of cabinets, cutout boxes, and meter socket enclosures. The following sections are among the key provisions of this article.

312.2 Damp and Wet Locations requires that all enclosures be designed and installed so as to prevent moisture or water from entering and/or accumulating within the cabinet, cutout box, or meter socket enclosure. **See Figure 9-10.**

312.6 Deflection of Conductors requires that sufficient space be provided for bending and installation of conductors.

312.8 Switch and Overcurrent Device Enclosures with Splices, Taps, and Feed-Through Conductors prohibits enclosures for overcurrent devices from being used for other purposes unless adequate space has been provided.

Article 312 is divided into two parts as follows:

I. Installation
- Damp, Wet, or Hazardous Locations
- Position in Wall
- Repairing Plaster and Drywall or Plasterboard
- Openings to be Closed
- Deflection of Conductors
- Space in Enclosures
- Enclosures for Switches or Overcurrent Devices
- Side or Back Wiring Spaces or Gutters

II. Construction Specifications
- Material
- Spacing

Article 314 Outlet, Device, Pull, and Junction Boxes; Conduit Bodies; Fittings; and Handhole Enclosures

Article 314 provides general rules for the installation and construction of outlet, device, pull and junction boxes; conduit bodies; fittings; and handhole enclosures. **See Figure 9-11.**

Figure 9-11   Handhole

Figure 9-11. Handhole boxes are utilized outdoors to provide pulling points and junction points for feeders and branch circuits. They are commonly found in landscaping, outdoor sports fields, and outdoor places where distances are great from the building and the outdoor equipment being served.

Figure 9-10   Meter Enclosure

Figure 9-10. Meter sockets have to be rated for damp and wet locations due to the weather elements.

For additional information, visit qr.njatcdb.org
Item #1071

Key provisions of **Article 314** include the following:

314.16 Number of Conductors in Outlet, Device, and Junction Boxes, and Conduit Bodies, along with the *Code* table provided, requires that all boxes and conduit bodies be of sufficient size to provide free space for enclosed conductors. Note that this section covers installations wherein all conductors enclosed are 6 AWG or smaller.

314.23 Supports provides minimum requirements for the support of all boxes and enclosures. Careful review of this section is important since its requirements are utilized in nearly every electrical installation.

314.28 Pull and Junction Boxes and Conduit Bodies provides minimum requirements for the size of all pull and junction boxes and conduit bodies. Note that this section covers installations wherein conductors 4 AWG or larger are enclosed.

314.29 Boxes, Conduit Bodies, and Handhole Enclosures to Be Accessible prohibits boxes and conduit bodies from being concealed. Note that a box installed above a lay-in type drop ceiling would be considered accessible.

314.71 Size of Pull and Junction Boxes, Conduit Bodies, and Handhole Enclosures is found in **Part IV** of **Article 314 Pull and Junction Boxes, Conduit Bodies, and Handhole Enclosures for Use on Systems over 600 Volts, Nominal**. It requires that all pull and junction boxes containing conductors over 600 volts be sized larger to accommodate the bending radius of high-voltage cables.

Article 314 is divided into four logical parts as follows:

I. Scope and General
- Round Boxes
- Nonmetallic Boxes
- Metal Boxes

II. Installation
- Damp or Wet Locations
- Number of Conductors in Outlet, Device, and Junction Boxes and Conduit Bodies
- Conductors Entering Boxes/Fittings and Conduit Bodies
- Boxes Enclosing Flush Devices
- Boxes in Walls or Ceilings
- Repairing Plaster, Drywall, or Plasterboard
- Exposed Surface Extensions
- Supports
- Depth of Outlet Boxes
- Covers and Canopies
- Outlet Boxes
- Size of Pull and Junction Boxes and Conduit Bodies
- Accessibility of Pull and Junction Boxes, Conduit Bodies, and Handhole Enclosures
- Handhole Enclosures

III. Construction Specifications
- Conduit Bodies, Metal Boxes, Fittings, and Covers
- Bushings
- Nonmetallic Boxes
- Marking

IV. Pull and Junction Boxes, Conduit Bodies, and Handhole Enclosures for Use on Systems over 600 Volts, Nominal
- General
- Size of Pull and Junction Boxes, Conduit Bodies and Handhole Enclosures
- Construction and Installation Requirements

CABLE ASSEMBLIES AND RACEWAYS

Similar Article Layout for Usability

All cable assembly and circular raceway articles share a common article layout and section numbering system intended to provide the *Code* user with a consistent, easy-to-use format. Many of the "other than circular" raceways have also adopted this common format. Once familiar with the common format of these articles, the *Code* user can quickly and accurately move through the many different wiring methods in Chapter 3 to find the needed information and requirements. For example, **Section XXX.10** of each article is "Uses Permitted." A *Code* user familiar with this common numbering scheme could quickly and accurately determine the wiring methods permitted for a particular installation.

Within the common format is included specific section numbering which may not apply to all wiring methods. For example, **Section 3XX.28** is reserved for "Reaming and Threading" and can be found only in metal raceway articles, where reaming and threading are necessary for installation.

The common article layout and section numbering is as follows:

Article 3XX
I. General
3XX.1 Scope
3XX.2 Definition(s)
3XX.3 Other Articles
3XX.6 Listing Requirements
II. Installation
3XX.10 Uses Permitted
3XX.12 Uses Not Permitted
3XX.15 Exposed Work
3XX.17 Through or Parallel to Framing Members
3XX.19 Clearances
3XX.20 Size
3XX.23 In Accessible Attics
3XX.24 Bends - How Made
3XX.26 Bends - Number in One Run
3XX.28 Reaming and Threading
3XX.30 Securing and Supporting
3XX.40 Boxes and Fittings
3XX.42 Couplings, Connectors, Devices
3XX.46 Bushings
3XX.48 Joints
3XX.56 Splices and Taps
3XX.60 Grounding, Bonding
3XX.80 Ampacity
III. Construction Specifications
3XX.100 Construction
3XX.104 Conductors
3XX.108 Equipment Grounding Conductor
3XX.112 Insulation
3XX.116 Sheath, Jacket, Conduit
3XX.120 Marking(s)
3XX.130 Standard Lengths

Cable Assemblies

Chapter 3 recognizes 11 types of cable assemblies as acceptable wiring methods. All cable assembly articles are listed in alphabetical order as follows:

Article 320 Armored Cable: Type AC
Article 322 Flat Cable Assemblies: Type FC
Article 324 Flat Conductor Cable: Type FCC
Article 326 Integrated Gas Spacer Cable: Type IGS
Article 328 Medium Voltage Cable: Type MV
Article 330 Metal-Clad Cable: Type MC
Article 332 Mineral-Insulated, Metal-Sheathed Cable: Type MI
Article 334 Nonmetallic-Sheathed Cable: Types NM, NMC, and NMS
Article 336 Power and Control Tray Cable: Type TC
Article 338 Service-Entrance Cable: Types SE and USE
Article 340 Underground Feeder and Branch-Circuit Cable: Type UF

Raceways, Circular Metal Conduit

Four types of circular metal conduits are recognized as an acceptable wiring method in Chapter 3. They are defined as rigid-type conduits or flexible-type conduits as follows:

Article 342 Intermediate Metal Conduit: Type IMC
Article 344 Rigid Metal Conduit: Type RMC

RMC includes Galvanized Steel, Stainless Steel, Brass, Aluminum, and Ferrous.

Figure 9-12   FMC (Article 348)

Figure 9-12. Flexible metal conduit can be used to connect electrical equipment that vibrates, or to connect equipment such as light fixtures in lay-in ceilings.

Article 348 Flexible Metal Conduit: Type FMC. See Figure 9-12.

Article 350 Liquidtight Flexible Metal Conduit: Type LFMC

Raceways, Circular Nonmetallic Conduit

Four types of circular nonmetallic conduits are recognized as an acceptable wiring method in Chapter 3. They are defined as rigid-type conduits or flexible-type conduits as follows:

Article 352 Rigid Polyvinyl Chloride Conduit: Type PVC

Article 353 High-Density Polyethylene Conduit: Type HDPE Conduit

Article 354 Nonmetallic Underground Conduit with Conductors: Type NUCC

Article 355 Reinforced Thermosetting Resin Conduit: Type RTRC

Article 356 Liquidtight Flexible Nonmetallic Conduit: Type LFNC

Raceways, Circular Metallic Tubing

Chapter 3 recognizes two types of circular metallic tubing as acceptable wiring methods. Note that electrical metallic tubing, type EMT, is commonly referred to as "thin-wall conduit" but is designated as "tubing" in the *NEC*. The two types of tubing, a single rigid type and a single flexible type, are covered in the following articles:

Article 358 Electrical Metallic Tubing: Type EMT

Article 360 Flexible Metallic Tubing: Type FMT

Raceways, Circular Nonmetallic Tubing

Chapter 3 recognizes a single type of circular nonmetallic tubing as an acceptable wiring method:

Article 362 Electrical Nonmetallic Tubing: Type ENT

Factory-Assembled Power Distribution Systems

The *NEC* recognizes two types of factory-assembled power distribution systems as acceptable wiring methods in Chapter 3. These wiring methods, busway and cablebus, are preassembled and bolted together for a complete installation in the field. These systems allow for a disconnecting means and overcurrent protection to be installed anywhere along the installation of busway and cablebus, providing an easy means for power distribution. The two articles for busway and cablebus numerically separate the "Circular Raceways" from the "Other than Circular Raceways," as follows:

Article 368 Busways

Article 370 Cablebus

> **Fact**
>
> Either torque tools are required for busway assembles or break-away bolt heads.

Raceways Other Than Circular

Ten types of other than circular raceways are recognized as an acceptable wiring method in Chapter 3:

Article 366 Auxiliary Gutters

Article 372 Cellular Concrete Floor Raceways

Article 374 Cellular Metal Floor Raceways

Article 376 Metal Wireways

Article 378 Nonmetallic Wireways

Article 380 Multioutlet Assembly
Article 384 Strut-Type Channel Raceway
Article 386 Surface Metal Raceways
Article 388 Surface Nonmetallic Raceways
Article 390 Underfloor Raceways

Surface-Mounted Nonmetallic Branch-Circuit Extension

This wiring method is primarily limited to use only from an existing outlet in a residential or commercial occupancy not more than three floors above grade. It was used primarily in older electrical installations to allow for the surface mounting of additional receptacle outlets. The 2008 *NEC* included provisions to allow a "concealable nonmetallic extension" to be installed on walls or ceilings covered with paneling, tile, joint compound, or similar material. This wiring method is covered in **Article 382 Nonmetallic Extensions.**

Support Systems for Cables/ Raceways

Two types of systems are recognized as acceptable for the support of wiring methods covered in Chapter 3:

Article 392 Cable Trays permits, under specified conditions, cable tray to support single conductors, cable assemblies, and raceways.

Article 396 Messenger-Supported Wiring permits, as the name implies, an exposed wiring support system using a messenger wire to support insulated conductors. This system is permitted to support only those cable assemblies or conductors listed in **Table 396.10(A)**.

Open-Type Wiring Methods

Three types of open-type wiring methods are permitted in Chapter 3, although they are of extremely limited use:

Article 394 Concealed Knob-and-Tube Wiring
Article 398 Open Wiring on Insulators
Article 399 Outdoor Overhead Conductors over 600 Volts

KEY WORDS AND CLUES FOR CHAPTER 3, "BUILD"

Wiring Methods
- General questions for wiring method installation
- Conductors, types, uses, ampacity
- Cable assemblies, all types
- Conduits, all types
- Tubing, all types
- Other raceways, all types
- Installation of all wiring methods
- Support of all wiring methods
- Construction of all wiring methods
- Uses permitted or not permitted, all wiring methods

Wiring Materials
- General questions for wiring materials
- Cabinets
- Cutout boxes
- Meter socket enclosures
- Outlet, device, pull and junction boxes
- Conduit bodies
- Handhole enclosures
- Support systems, cable tray, and messenger-supported wiring
- Construction of wiring materials
- Installation of wiring materials
- Support of wiring materials

Think "Build" and Go to Chapter 3
Refer to Chapter 3 when your question or need within the *NEC* deals with any of the following:
- Wiring methods, cable assemblies, raceways
- Conductors, type designation, use, ampacity
- Installation of wiring methods
- Wiring materials, enclosures of all types
- Installation, use, and size of wiring materials
- Any question on the physical installation of wiring methods and materials
- Support systems for raceways, conductors, or cable assemblies

Summary

Chapter 3, in accordance with **90.3**, applies generally to all electrical installations. Using the *Codeology* method, this chapter has been coined the "Build" chapter due to the scope of its coverage. Chapter 3 is a "hands-on" and mechanical chapter, in that all of the material covered is to be physically installed. It is the means by which electrical current is delivered from the source of power to the last outlet in the electrical distribution system.

From the *NEC* title for **Chapter 3 Wiring Methods and Materials**, the scope of this chapter can be described as "Information and Rules on Wiring Methods and Materials for Electrical Installations." Chapter 3 covers the entire electrical distribution system, from the service point (connection to the utility) to the last receptacle or other outlet in the electrical system. All of the wiring methods and materials used to distribute electrical energy from the source to the last outlet are covered in Chapter 3. In accordance with its scope of "Wiring Methods," Chapter 3 provides detailed requirements for all conductors, raceways, cable assemblies, and other recognized wiring methods. In accordance with its scope of "Wiring Materials," it provides detailed requirements for all enclosures, boxes, conduit bodies, and support systems.

Chapter 3, along with Chapters 1, 2, and 4, builds the foundation or backbone of all electrical installations. **See Figure 9-13.**

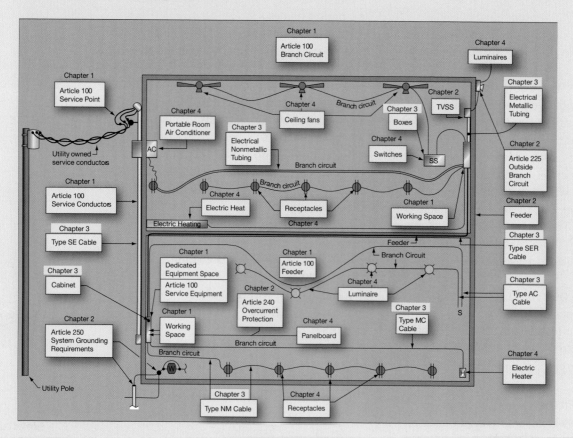

Figure 9-13. Chapter 3 applies generally in all electrical installations.

Review Questions

1. Does Chapter 3 of the *NEC* apply to all electrical installations covered by the *NEC*?

2. Chapter 3 is subdivided into how many articles?

3. The scope of *NEC* Chapter 3 is divided into two areas covering wiring __?__ and wiring __?__.

4. Name the part and article of Chapter 3 which addresses the installation of a pull box containing conductors rated at 13,200 volts.

5. Chapter 3 of the *NEC* addresses wiring methods, which would include circular raceways. Name the part and article which would apply to the installation of type RMC conduit.

6. Does 300.22(C) provide any information to aid the *Code* user in understanding what "other space" is in relation to a duct or plenum?

7. Which two articles in *NEC* Chapter 3 cover requirements for support systems for cables and/or raceways?

8. Article 340 is divided into how many parts?

9. Does Part III of Article 340 apply to the installation of type UF cable?

10. Which two articles in Chapter 3 cover requirements for factory-assembled power distribution systems?

To access Practice
Problems, visit
qr.njatcdb.org.
Item #1082
Click on Inside
Blended Learning.

Chapter 4 of the *NEC*, "USE"

C hapter 4 of the *NEC*, the "Use" chapter, provides rules and information on electrical equipment for general use. Special equipment is addressed in Chapter 6, in accordance with **Section 90.3**.

Chapter 2 of the *NEC* "Plans" for a general electrical installation. Chapter 3 "Builds" a general electrical installation by delivering electrical energy from the source to the load(s). All equipment covered in Chapter 4 is dedicated to the "Use" of electrical energy. Chapter 4 addresses the use or consumption of electrical energy including utilization and all other necessary equipment. Electrical equipment, which uses electrical energy, performs a task or provides a service for the consumer. For example, lighting fixtures illuminate our homes, electrical space heaters provide heat, and air conditioners cool our homes.

Objectives

» Associate the *Codeology* title for *NEC* Chapter 4 as "Use."

» Identify the type of information and requirements dealing with the installation, control, and supply for utilization equipment covered in *NEC* Chapter 4.

» Recognize Chapter 4 numbering as the 400-series.

» Recognize, recall, and become familiar with articles contained in Chapter 4.

Table of Contents

NEC CHAPTER 4, "USE"

While Chapter 4 of the *NEC* is identified as the "Use" chapter, not all of the equipment it covers uses electrical energy. However, all of the equipment in Chapter 4 plays a major role in the use of electrical energy. The following examples illustrate the makeup of Chapter 4, the "Use" chapter:

- Luminaires (lighting fixtures), appliances, heating equipment, motors, and air-conditioning/refrigeration equipment all use electrical energy.
- Flexible cords and cables allow for the connection of appliances and other utilization equipment to an electrical outlet.
- Fixture wires provide for the wiring of luminaires.
- Panelboards, switchboards, industrial control panels, and switches provide control and overcurrent protection for all conductors supplying end-use equipment.
- Receptacle outlets facilitate the use of appliances and other loads.
- Switches control lighting and all other loads.
- Generators provide a power source for emergency, legally required standby, and optional standby systems. Other sources of power, such as solar, wind, and fuel cell systems are considered special and are covered by Chapter 6 of the *NEC*.
- Transformers enable the use of electrical energy and provide the flexibility to create a new system to allow the consumption of electrical energy at utilization voltages. For example, a service may be 277/480 volts, 3-phase, 4-wire to supply air-conditioning and refrigeration equipment. A transformer is installed which creates a new system at 120/208 volts to allow electrical energy consumption at 120 volts for general receptacle outlets.
- Phase converters, capacitors, resistors, and reactors allow for the economical use of electrical energy.

The 21 articles in Chapter 4, the 400-series, address equipment for general use to facilitate the utilization of electrical energy. The requirements and information in this chapter are logically divided into seven categories. **See Figure 10-1.**

CATEGORIZATION OF CHAPTER 4

When using the *Codeology* method, the title for Chapter 4 is "Use." This title encompasses all equipment which uses electrical energy as well as all associated equipment necessary to safely accomplish utilization. To further explain, equipment which uses energy includes luminaires (light fixtures), appliances, heaters, motors, and air-conditioning units. Auxiliary equipment associated with motors and transformers are capacitors, resistors, and reactors. Equipment which safely accomplishes utilization (distributes electrical energy) include cords, fixture wires, receptacles, switches, switchboards, panelboards, transformers, and phase converters. Additional equipment in Chapter 4 which also provides utilization of electricity is generators and batteries; which produce energy. **See Figure 10-2.**

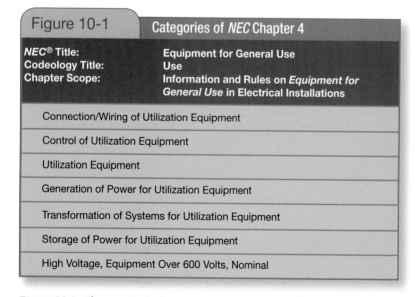

Figure 10-1	Categories of *NEC* Chapter 4
NEC® Title: **Codeology Title:** **Chapter Scope:**	**Equipment for General Use** **Use** **Information and Rules on *Equipment for General Use* in Electrical Installations**
Connection/Wiring of Utilization Equipment	
Control of Utilization Equipment	
Utilization Equipment	
Generation of Power for Utilization Equipment	
Transformation of Systems for Utilization Equipment	
Storage of Power for Utilization Equipment	
High Voltage, Equipment Over 600 Volts, Nominal	

Figure 10-1. Chapter 4 is broken into seven categories facilitating the utilization of electrical energy.

Figure 10-2   **Equipment for General Use**

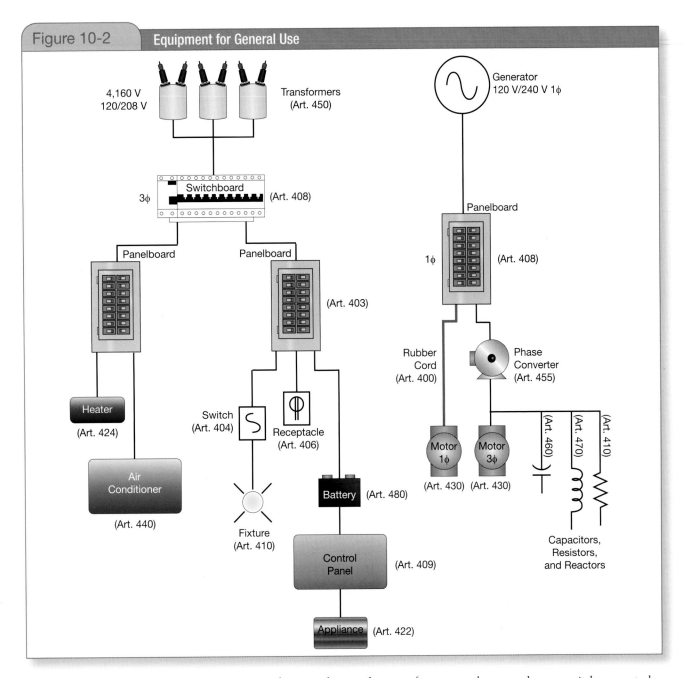

Figure 10-2. Chapter 4 of the NEC covers Equipment for General Use such as transformers, panels, motors, heaters, switches, receptacles, and flexible cords.

Connection and Wiring of Utilization Equipment

Flexible cords and cables are appropriately located in Chapter 4, the "Use" chapter. Flexible cords and cables are not permitted as a substitute for fixed wiring. This means that while cords and cables are permitted to supply utilization equipment, they are not permitted as "wiring methods." Chapter 3 of the *NEC* is dedicated to wiring methods and wiring materials.

Fact

The typical use for flexible cords and cables include the wiring of pendants/luminaires, elevators, and cranes/hoists, and reducing the transmission of noise and vibrations.

Flexible cord and cables facilitate the installation of equipment which is frequently changed or partially mobile. **See Figure 10-3.**

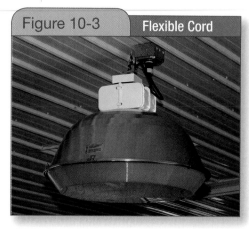

Figure 10-3   Flexible Cord

Figure 10-3. Typically, luminaires are supplied with power by flexible cords and cables.

The *NEC* uses the widely known international term for lighting fixtures, *luminaires*. **Article 402** provides information and requirements for the use and limits of fixture wire in luminaires and associated equipment. It is important to note that fixture wires are not permitted to serve as branch circuits. They are permitted only for installation in lighting fixtures or associated equipment. As covered in **Article 310**, fixture wires are rated at higher temperatures than normal general use conductors due to the heat produced in most luminaire fixtures.

Electrical systems are installed in all occupancies to allow for the use of electrical equipment. Utilization equipment must be connected to a branch circuit through hardwiring or cord-and-plug connection. **Article 400** and **Article 402** provide information and requirements on the use of cords and cables and fixture wires.

Article 400 Flexible Cords and Cables

I.   **General**
II.   **Construction Specifications**
III.   **Portable Cables Over 600 Volts, Nominal**

Article 402 Fixture Wires

Control of Utilization Equipment

Switches are appropriately located in Chapter 4, the "Use" chapter. Switches are essential to the control and protection (fused switches) of conductors and utilization equipment. Receptacles, cord connectors, and attachment plugs are also covered by Chapter 4, the "Use" chapter. The connection and disconnection of appliances and other cord-and-plug-connected utilization equipment would not be possible without these devices. In addition, switchboards, panelboards, and industrial control panels are also covered by Chapter 4 and play an essential part to allow for control and protection of all conductors and utilization equipment.

The *NEC* requires control and protection of all conductors and equipment in an electrical installation. In some cases, utilization equipment is controlled and protected through the same devices used for overcurrent protection of branch-circuit conductors. However, not all electrical "equipment for general use" is located at the end of the branch circuit with the utilization equipment. **Article 408 Switchboards and Panelboards** and **Article 409 Industrial Control Panels** provide provisions for the control and protection of utilization equipment. In most cases, this occurs at the source of the feeder and branch circuits supplying the equipment. Switchboards typically provide distribution to larger loads and other panels. **See Figure 10-4a.** Panelboards are typically full of 15- and 20-amp breakers feeding branch circuits of receptacles and luminaires. **See Figure 10-4b. Article 404 Switches** may cover only the utilization equipment or part of a branch circuit, while **Article 406 Receptacles, Cord Connectors, and Attachment Plugs (Caps)** covers the utilization equipment providing control and flexibility. **See Figure 10-4c.**

Figure 10-4a. *Switchboards typically supply service to larger loads such as motors, transformers, and larger ampacity equipment.*

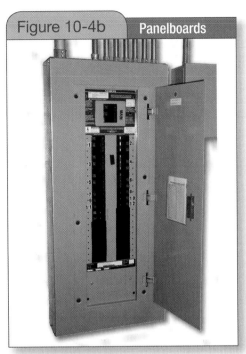

Figure 10-4b. *Panelboards are typically used for branch circuits used to supply wiring devices such as receptacles.*

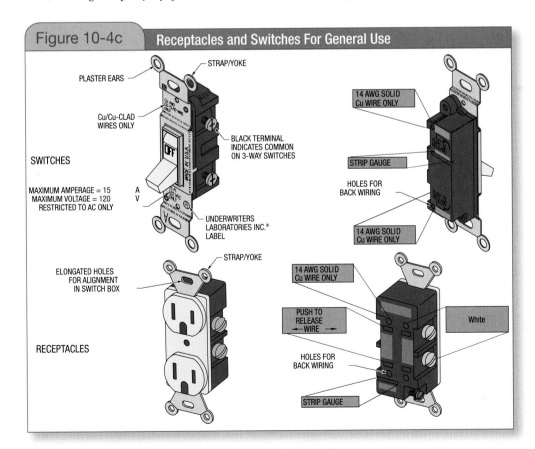

Figure 10-4c. *Receptacles and switches are equipment for general use.*

Four articles address the control requirements of Chapter 4:

Article 404 Switches
I.   **Installation**
II.   **Construction Specifications**

Article 406 Receptacles, Cord Connectors, and Attachment Plugs (Caps)

Article 408 Switchboards and Panelboards
I.   **General**
II.   **Switchboards**

> **Switchboard.** A large single panel, frame, or assembly of panels on which are mounted on the face, back, or both, switches, overcurrent and other protective devices, buses, and usually instruments. Switchboards are generally accessible from the rear as well as from the front and are not intended to be installed in cabinets.

III. **Panelboards**

> **Panelboard.** A single panel or group of panel units designed for assembly in the form of a single panel, including buses and automatic overcurrent devices, and equipped with or without switches for control of light, heat, or power circuits; designed to be placed in a cabinet or cutout box placed in or against a wall, partition, or other support; and accessible only from the front.

IV.   **Construction Specifications**

Article 409 Industrial Control Panels
I.   **General**
II.   **Installation**
III. **Construction Specifications**

Categories of Utilization Equipment

Luminaires (lighting fixtures), appliances, space-heating equipment, deicing and snow-melting equipment, heat trace, motors, AC, and refrigeration equipment are all appropriately located in Chapter 4, the "Use" chapter. This equipment uses electrical energy. **See Figure 10-5.**

Eight articles in the *NEC* specifically address the requirements for utilization equipment. These requirements include those for general coverage, installation, control and protection, disconnecting means, construction, and many other equipment-specific needs. The *NEC* requirements for special utilization equipment are located in Chapter 6, in conformance with **Section 90.3**. General utilization equipment in Chapter 4 of the *NEC* is divided into four basic categories: lighting, appliances, heating, and motors.

LIGHTING

Article 410 Luminaires, Lampholders, and Lamps
I.   **General**
II.   **Luminaire Locations**
III. **Provisions at Luminaire Outlet Boxes, Canopies, and Pans**
IV.   **Luminaire Supports**
V.   **Grounding**

Figure 10-5a. Article 410 covers the installation of luminaires, while Article 422 spells out the requirements for the typical appliances such as a ceiling fan.

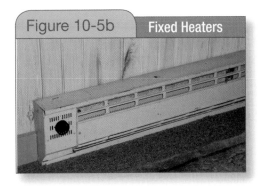

Figure 10-5b. Article 424 provides details of installation for fixed space heating equipment.

Figure 10-5c. Article 430 lists the installation requirements of motors.

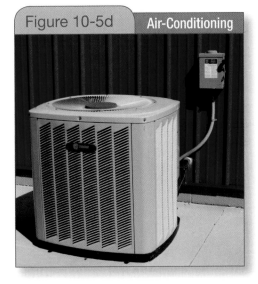

Figure 10-5d. Article 440 covers the installation of air-conditioning and refrigerating equipment.

Figure 10-6   Generator

Figure 10-6. Generators are necessary in many types of installations to supply power for utilization equipment. Note that this generator is mounted on top of the day fuel tank.

Generation of Power for Utilization Equipment

Coverage of generators is appropriately located in Chapter 4, the "Use" chapter. In almost all electrical installations, the power is supplied to an occupancy by an electric utility and is called a service. When the need for an emergency or standby system arises, the most common or general backup system is an on-site standby generator. **See Figure 10-6.**

Article 445 provides information and requirements for generators, which are considered equipment for general use. In accordance with **Section 90.3**, the *NEC* addresses two additional "special" energy systems in Chapter 6. **Article 445 Generators** addresses the generation requirements of Chapter 4 and is not divided into parts.

Figure 10-7a. *Pad-mounted transformers typically feed the entire building with power.*

Transformation of Systems for Utilization Equipment

Transformers are appropriately located in Chapter 4, the "Use" chapter. When utilization equipment operates at different system voltages in an electrical installation, transformers are used to derive new systems at the utilization voltage to meet the system requirements. **See Figure 10-7.**

Phase converters, capacitors, resistors, and reactors are also covered in Chapter 4 and allow economical and efficient use of utilization equipment. For example, consider an older building or structure with a 2-phase, 5-wire service. To use a 3-phase motor in this occupancy, which is much less expensive, a phase converter would be required. In addition, power factor problems of wasted energy and high cost can be corrected through the application of capacitors.

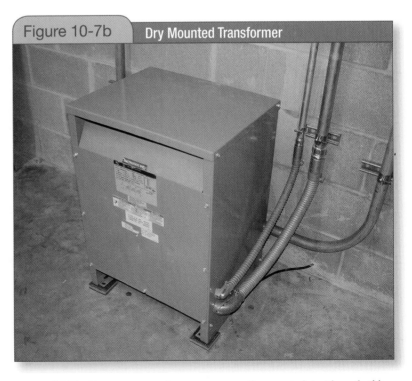

Figure 10-7b. *Dry type transformers are usually mounted inside a building providing power for specific loads or an additional voltage service.*

Article 450 covers transformers, which are an essential part of most electrical installations. They provide an installation with the flexibility of deriving a new system to meet the requirements of utilization equipment. Phase converters, covered by **Article 455**, allow single-phase systems to derive a 3-phase system and older 2-phase electrical systems to utilize the 3-phase equipment, which is more common, less expensive, and readily available. While capacitors, resistors, and reactors do not transform or change a system, they are essential equipment which supplements electrical systems in special applications.

Four articles address the transformation or adjustment of system requirements of Chapter 4:

Article 450 Transformers and Transformer
Vaults (Including Secondary Ties)
I.   **General Provisions**
II.   **Specific Provisions Applicable to**
      Different Types of Transformers
III. **Transformer Vaults**

Figure 10-8   Battery Rack

Figure 10-8. Batteries are equipment for general use which store energy for utilization equipment.

Article 455 Phase Converters
I.   **General**
II.   **Specific Provisions Applicable to**
      Different Types of Phase Converters

> **Fact**
>
> There are primarily two types of phase converters: Static and Rotary. Static units can only supply motors, whereas Rotary units can serve both motors and resistive loads.

Article 460 Capacitors
I.   **600 Volts, Nominal, and Under**
II.   **Over 600 Volts, Nominal**

Article 470 Resistors and Reactors
I.   **600 Volts, Nominal, and Under**
II.   **Over 600 Volts, Nominal**

Storage of Power for Utilization Equipment

Batteries are appropriately located in Chapter 4, the "Use" chapter. **See Figure 10-8.**

Storage batteries are used in a variety of electrical installations to provide power for utilization equipment in the event of a power loss. **Article 480 Storage Batteries** addresses the energy storage requirements of Chapter 4 and provides information and requirements for the safe installation of these battery-supplied systems. **Article 480** is not divided into separate parts.

> **Fact**
>
> Article 480.9 requires sufficient diffusion and ventilation of the gases from the battery to prevent the accumulation of an explosive mixture. This article also requires battery locations to be in compliance with 110.26 Working Space and 110.27 Live Parts.

Equipment Over 600 Volts, Nominal

The specific requirements necessary for electrical equipment for general use, rated at over 600 volts nominal, are addressed by **Article 490**. Often, distribution voltages range between 4,160 to 13,800 volts and require special installation details. The equipment can be indoor rated only or installed outside. **See Figure 10-9.**

Article 490 Equipment, Over 600 Volts, Nominal

I.   **General**
II. **Equipment—Specific Provisions**
III. **Equipment—Metal-Enclosed Power Switchgear and Industrial Control Assemblies**
IV. **Mobile and Portable Equipment**
V.   **Electrode-Type Boilers**

KEY WORDS AND CLUES FOR CHAPTER 4, "USE"

The control, connection, installation, protection, or construction specifications of general utilization equipment include the following:

- Luminaires (lighting fixtures)
- Fixture wire
- Cords and cables
- Receptacles
- Cord connectors, attachment plugs
- Switchboards
- Panelboards
- Industrial control panels
- Low-voltage lighting
- Appliances
- Fixed electric space-heating equipment
- Fixed outdoor electric deicing and snow-melting equipment

Figure 10-9 | **Switchgear**

Figure 10-9. Typically, a transformer is mounted near the high voltage switchgear. Article 490 covers details for installations over 600 volts.

- Fixed electric heating equipment for pipelines and vessels (heat trace)
- Motors, motor circuits, and controllers
- Air-conditioning and refrigeration equipment
- Generators
- Transformers and transformer vaults
- Phase converters
- Capacitors
- Resistors and reactors
- Storage batteries
- Electrical equipment, over 600 volts

Summary

In accordance with **Section 90.3**, *NEC* Chapter 4 applies generally to all electrical installations. Using the *Codeology* method, this chapter is coined the "Use" chapter due to the scope of its coverage. All of the material covered pertains to "electrical equipment for general use." The scope of this chapter is directed at equipment which uses electrical energy and associated equipment necessary for safe utilization.

Chapter 4 of the *NEC* is titled "Equipment for General Use," from which its scope can be described as "Information and Rules on Equipment for General Use in Electrical Installations." Chapter 4 covers the entire electrical distribution system from the service point (connection to the utility) to the last receptacle or other outlet in the electrical system for electrical equipment. Wiring methods and wiring materials are not covered here. They are found in Chapter 3.

Chapter 4, along with Chapters 1, 2, and 3, build the foundation or backbone of all electrical installations. **See Figure 10-10.**

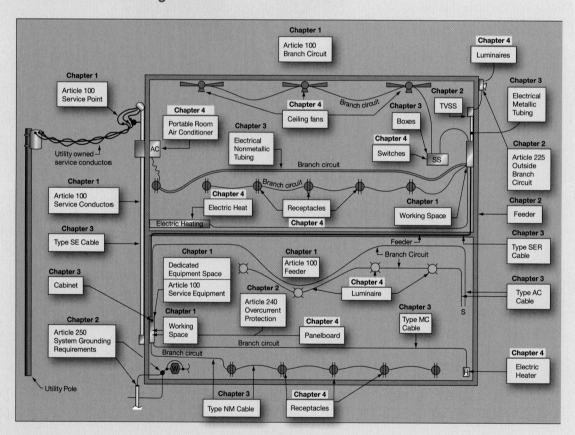

Figure 10-10. Chapter 4 will apply generally to all electrical installations.

Review Questions

1. Does Chapter 4 of the *NEC* apply to all electrical installations covered by the *Code*?

2. Chapter 4 is subdivided into how many articles?

3. The scope of Chapter 4 of the *NEC* is dedicated to equipment for general use to facilitate the __?__ of electrical energy.

4. What part of which article in Chapter 4 would address the installation of track lighting?

5. Chapter 4 of the *NEC* addresses equipment for general use, which would include motors. What part of which article would apply if one were sizing motor overload protection?

6. Does Section 426.20 apply to resistance heating elements?

7. Which four articles in Chapter 4 cover requirements for the control of utilization equipment?

8. Into how many parts is Article 490 subdivided?

9. Does Part III of Article 440 apply to the installation of portable room air conditioners?

10. Which article in Chapter 4 covers requirements for generation of power for utilization equipment?

To access Practice Problems, visit qr.njatcdb.org. Item #1082 Click on Inside Blended Learning.

Chapters 5, 6, and 7 of the *NEC*, "SPECIAL"

In accordance with **Section 90.3** of the *NEC*, Chapters 1 through 4 are general in scope and provide installation requirements for an entire electrical system. Chapters 5, 6, and 7 are the applications of special requirements, supplements, or modifications to the first four chapters. All of the requirements in these "Special" chapters typically modify the basic rules or apply supplemental requirements to address special needs. The "Special" chapters contain 63 articles, making it necessary to become familiar with the different types of special occupancies, equipment, and conditions.

Objectives

» Associate the *Codeology* title for *NEC* Chapter 5 as "Special Occupancies," Chapter 6 as "Special Equipment," and Chapter 7 as "Special Conditions."

» Identify the special type of information and requirements contained in Chapters 5, 6, and 7 for supplementing and/or modifying the requirements of Chapters 1 through 4.

» Recognize Chapters 5, 6, and 7 as the 500-, 600-, and 700-series.

» Recognize, recall, and become familiar with articles contained in Chapters 5, 6, and 7.

Chapter 11

Table of Contents

STRUCTURE OF THE "SPECIAL" CHAPTERS

The need to address modifications and supplemental requirements covered by the "Special" chapters must be identified before any work is started on an electrical installation. Each stage of an electrical installation is affected when a special situation is encountered. Therefore, the special requirements of Chapters 5, 6, and 7 must be considered in each step of the installation to prevent serious misapplication of the *NEC*.

For example, a new hospital is to be constructed and the electrical installation is being studied. A hospital is considered a special occupancy because of the specific requirements for uninterrupted power, protection of patients, and many other special needs. The *NEC* covers the special needs of a hospital in **Article 517 Health Care Facilities. See Figure 11-1.**

PLANNING STAGES MODIFIED AND/OR SUPPLEMENTED BY CHAPTERS 5, 6, AND 7

Continuing our example of planning for new hospital construction, the requirements of **Chapter 5 Special Occupancies**, will affect the Plan stage of the hospital. For example, **Article 517** will modify and supplement the Plan stage of the electrical installation through special requirements for grounding, branch circuits, and receptacle locations for patient care areas. These special provisions require specific wiring methods to the patient care areas, thereby affecting the Build stage of this installation. The Use stage of this installation will be modified and supplemented by requirements, including those for therapeutic pools and tubs, in **Chapter 6 Special Equipment.** All stages of the installation will be further modified and supplemented by **Chapter 7 Special Conditions,** when emergency systems are installed to meet life safety requirements.

NEC CHAPTER 5, "SPECIAL OCCUPANCIES"

The *Codeology* title for *NEC* Chapter 5 is "Special Occupancies." All stages of an electrical installation will be modified or supplemented when special occupancies are involved. The 28 articles of Chapter 5 can be grouped as follows:

HAZARDOUS LOCATIONS
Article 500 Hazardous (Classified)
            Locations, Classes I, II, and III,
            Divisions 1 and 2
Article 501 Class I Locations
Article 502 Class II Locations
Article 503 Class III Locations
Article 504 Intrinsically Safe Systems

ZONE CLASSIFICATION SYSTEM
Article 505 Zone 0, 1, and 2 Locations
Article 506 Zone 20, 21, and 22 Locations
            for Combustible Dusts or
            Ignitable Fibers/Flyings
Article 510 Hazardous (Classified) Locations
            - Specific
Article 511 Commercial Garages, Repair
            and Storage
Article 513 Aircraft Hangars
Article 514 Motor Fuel Dispensing Facilities
Article 515 Bulk Storage Plants
Article 516 Spray Application, Dipping and
            Coating Processes

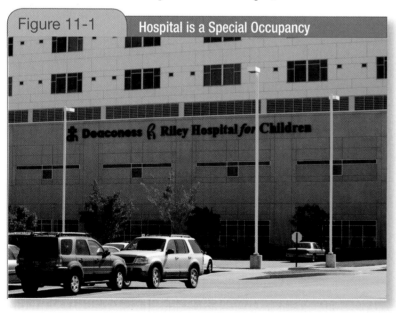

Figure 11-1   Hospital is a Special Occupancy

Figure 11-1. Chapter 5 covers special occupancies such as hospitals.

HEALTH CARE FACILITIES
Article 517 Health Care Facilities

ASSEMBLY OCCUPANCIES FOR 100 OR MORE PERSONS
Article 518 Assembly Occupancies

ENTERTAINMENT VENUES
Article 520 Theatres, Audience Areas of Motion Picture and Television Studios, Performance Areas, and Similar Locations
Article 522 Control Systems for Permanent Amusement Attractions
Article 525 Carnivals, Circuses, Fairs, and Similar Events
Article 530 Motion Picture and Television Studios and Similar Locations
Article 540 Motion Picture Projection Rooms

AGRICULTURAL BUILDINGS, POULTRY, LIVESTOCK, AND FISH
Article 547 Agricultural Buildings

MANUFACTURED BUILDINGS, DWELLINGS, AND RECREATIONAL STRUCTURES
Article 545 Manufactured Buildings
Article 550 Mobile Homes, Manufactured Homes, and Mobile Home Parks
Article 551 Recreational Vehicles and Recreational Vehicle Parks
Article 552 Park Trailers

STRUCTURES ON OR ADJACENT TO BODIES OF WATER
Article 553 Floating Buildings
Article 555 Marinas and Boatyards

TEMPORARY INSTALLATIONS
Article 590 Temporary Installations

Hazardous Locations

Article 500 through **Article 516** contain modifications and supplemental requirements for occupancies which contain, process, manufacture, or store materials which could cause a fire or explosion due to flammable gases or vapors, flammable liquids, combustible dust(s) or ignitable

Figure 11-2 | **Industrial Chemical Processing Plant**

Figure 11-2. Hazardous locations which process ignitable materials, liquids, and vapors are special occupancies.

fibers and flyings. **See Figure 11-2.**

A new *Code* user will often think of industrial locations as the only hazardous locations, but in reality, the general public typically visits one of the most common hazardous locations at least once a week. Portions of gas stations used to refuel vehicles fall under the classification of hazardous locations. Only where the gas is being dispensed is considered hazardous; whereas the convenient store associated with the typical gas station is not considered a hazardous location and is covered by Chapters 1 through 4 of the *NEC.* **See Figure 11-3.**

Article 500 sets the stage for the application of **Articles 501, 502, 503,** and **504. Section 500.2** contains definitions which apply to all hazardous location articles (**Article 500** through **Article 516**), except for the two zone classification articles (**Article 505** and **Article 506**), which provide an alternative to the Class I, II, and III systems. **Article 500** provides the basic information and requirements for the application of Class I, II, and III systems for hazardous locations. The following key provisions of **Article 500** apply to all hazardous location articles except for zone system **Articles 505** and **506.**

Figure 11-3   Gas Station with Convenience Store

Figure 11-3. Not all areas of a gas station are considered hazardous, such as the inside of the convenience store associated with the gas station.

Article 500 Hazardous (Classified) Locations, Classes I, II, and III, Divisions 1 and 2

500.2 Definitions

The following definitions apply to **Articles 501, 502, 503, 504, 510, 511, 513, 514, 515, and 516**:

- *Associated Nonincendive Field Wiring Apparatus*
- *Combustible Dust*
- *Combustible Gas Detection System*
- *Control Drawing*
- *Dust-Ignitionproof*
- *Dusttight Electrical and Electronic Equipment*
- *Explosionproof Equipment*
- *Hermetically Sealed*
- *Nonincendive Circuit*

- *Nonincendive Component*
- *Nonincendive Equipment*
- *Nonincendive Field Wiring*
- *Nonincendive Field Wiring Apparatus*
- *Oil Immersion*
- *Purged and Pressurized*
- *Unclassified Locations*

500.4 General
500.4(A) Documentation

500.4(A) requires that all areas designated as hazardous locations be properly documented and that the documentation be available to designers, installers, maintainers, and operators at the location.

500.4(B) Reference Standards

The five informational notes comprising 500.4(B) inform the *Code* user of other applicable codes necessary for proper application of the *NEC* requirements.

500.5 Classifications of Locations

Electrical installations in classification areas require electrical materials and equipment designed to prevent sparks from occurring around flammable vapors, liquids, gases, and dusts/fibers. These items are not used in basic commercial and industrial locations and may be unfamiliar to the installer. Items such as hazardous rated push button switches, receptacles, and light fixtures must be installed in classified areas. **See Figure 11-4.**

Not only does electrical equipment which produces sparks need to be hazardous rated, so does the conduit system.

For additional information, visit qr.njatcdb.org Item #1072

Figure 11-4   Hazardous Rated Equipment

Figure 11-4. Equipment which creates sparks when utilized must be designed to refrain from igniting flammable vapors, liquids, gases, and dusts/fibers.

Figure 11-5 **Hazardous Rated Raceway Components**

Figure 11-5. Equipment which creates sparks when utilized must be connected with a raceway system which is installed for hazardous areas.

The conduit system has to prevent ignitable vapors, liquids, gases, and dust/fibers from entering the raceway and being delivered to the equipment creating sparks. Junction boxes, flexible connections, and 90 degree pull boxes must all be hazardous rated. **See Figure 11-5.**

500.5(A) Classifications of Locations

The classification of hazardous locations is based upon two factors:

1. The properties of the flammable vapors, liquids, gases or flammable liquid-produced, combustible dusts or fibers which may be present

2. The likelihood that a flammable or combustible concentration or quantity is present

500.5(B) Class I Locations

Class I locations are those in which flammable gases or flammable liquid-produced vapors are or may be present in the air in quantities sufficient to produce explosive or ignitable mixtures. **See Figure 11–2.**

(1) Class I, Division 1

- Locations in which ignitable concentrations exist under normal operations
- Locations in which ignitable concentrations may exist due to repair, maintenance, or leaks
- Locations in which ignitable concentrations exist due to processes, breakdown, or faulty equipment

(2) Class I, Division 2

- Locations in which volatiles are handled, processed, or used in closed containers

- Locations in which positive ventilation prevents accumulation of gases/vapors
- Areas adjacent to Class I, Division 1 locations

500.5(C) Class II Locations

Class II locations are those that are hazardous because of the presence of combustible dust. An example of a Class II location is a storage grain bin which produces combustible dust during the movement of the grain. **See Figure 11-6.**

Other Class II hazardous location examples are flour and feed mills; producers of plastics, medicines, and fireworks; producers of starch or candies; spice-grinding plants; sugar plants; and cocoa plants.

For additional information, visit qr.njatcdb.org Item #1073

Figure 11-6   **Storage of Grain is a Hazardous Location**

Figure 11-6. Across the heartland of America, farmers utilize grain storage bins which must be hazardous rated. It makes no difference whether it is a couple of storage bins or a large distribution center; they are all considered hazardous.

(1) Class II, Division 1
- Locations in which ignitable concentrations of combustible dust exist under normal operations
- Where mechanical or machinery failure, repair, maintenance or leaks could create ignitable concentrations of dust
- Locations in which metal dusts exist, including, but not limited to, aluminum and magnesium

(2) Class II, Division 2
- Locations in which ignitable concentrations of combustible dust may exist due to abnormal operations
- Locations in which combustible dust is present, but not in sufficient amounts, unless a malfunction of equipment or process occurs
- Locations in which accumulating dust interferes with heat dissipation and/or could be ignited through equipment failure

500.5(D) Class III Locations

Class III locations are those that are hazardous because of the presence of easily ignitable fibers or flyings, but in which such fibers or flyings are not likely to be in suspension in the air in quantities sufficient to produce ignitable mixtures.

(1) Class III, Division 1
- Locations in which easily ignitable fibers or materials producing combustible flyings are handled, manufactured, or used

(2) Class III, Division 2
- Locations in which easily ignitable fibers are stored or handled other than in the process of manufacture

500.6 Material Groups

Section 500.6 covers the group classifications for equipment listings installed in areas where combustible gases, vapors, and dust:
- Class I, Group A, B, C, and D
- Class II Group E, F, and G

500.7 Protection Techniques

The protection techniques permitted for electrical and electronic equipment are listed in 11 first-level subdivisions.

500.8 Equipment

Minimum equipment requirements are listed and include the following:
- Suitability
- Approval
- Marking
- Temperature
- Threading

Article 501 Class I Locations

I. General
- Scope
- General Rules

II. Wiring
- Wiring Methods
- Sealing and Drainage
- Process Sealing
- Conductor Insulation
- Uninsulated Exposed Parts
- Grounding and Bonding
- Surge Protection

III. Equipment
- Transformers and Capacitors
- Meters, Instruments, and Relays
- Switches, Circuit Breakers, Motor Controllers, and Fuses
- Control Transformers and Resistors
- Motors and Generators
- Luminaires
- Utilization Equipment
- Flexible Cords
- Receptacles and Attachment Plugs
- Signaling, Alarm, Remote-Control, and Communications Systems

Article 502 Class II Locations

I. General
- Scope
- General Rules

II. Wiring
- Wiring Methods
- Sealing
- Uninsulated Exposed Parts
- Grounding and Bonding
- Surge Protection
- Multiwire Branch Circuits

III. Equipment
- Transformers and Capacitors
- Switches, Circuit Breakers, Motor Controllers, and Fuses

- Control Transformers and Resistors
- Motors and Generators
- Ventilating Piping
- Luminaires
- Utilization Equipment
- Flexible Cords
- Receptacles and Attachment Plugs
- Signaling, Alarm, Remote-Control, and Communications Systems; and Meters, Instruments, and Relays

Article 503 Class III Locations

I. General
- Scope
- General Rules

II. Wiring
- Wiring Methods
- Uninsulated Exposed Parts
- Grounding and Bonding

III. Equipment
- Transformers and Capacitors
- Switches, Circuit Breakers, Motor Controllers, and Fuses
- Control Transformers and Resistors
- Motors and Generators
- Ventilating Piping
- Luminaires
- Utilization Equipment
- Flexible Cords
- Receptacles and Attachment Plugs
- Signaling, Alarm, Remote-Control, and Local Loudspeaker Intercommunications Systems
- Electric Cranes, Hoists, and Similar Equipment
- Storage Battery Charging Equipment

Article 504 Intrinsically Safe Systems

Intrinsically safe circuit is defined in **Section 504.2** as a one in which "any spark or thermal effect is incapable of causing ignition of a mixture of flammable or combustible material in air under prescribed test conditions." Intrinsically safe systems are, by design, incapable of causing fire or explosion. These systems are designed and identified for use in hazardous locations. The function behind these systems is that the current and voltage is limited to the point where there is not enough electrical energy to produce a spark. Typically, these components are mounted outside the hazardous area and only the low-energy conductors enter the hazardous area. **See Figure 11-7.**

Zone Classification System

The Zone Classification System provides the *Code* user with an alternative to the Class I, II, and III systems outlined in **Articles 500, 501, 502,** and **503. Article 505** covers Class I locations and **Article 506** covers Class II and III locations.

Article 505 Zone 0, 1, and 2 Locations

Class I, Zone 0
- Ignitable concentrations are present continuously
- Ignitable concentrations are present for long periods of time

Class I, Zone 1
- Ignitable concentrations are likely exist under normal operations
- Ignitable concentrations may exist due to repair, maintenance, or leaks
- Ignitable concentrations exist due to processes, breakdown, or faulty equipment
- Areas adjacent to Class I, Zone 0 locations

For additional information, visit qr.njatcdb.org Item #1074

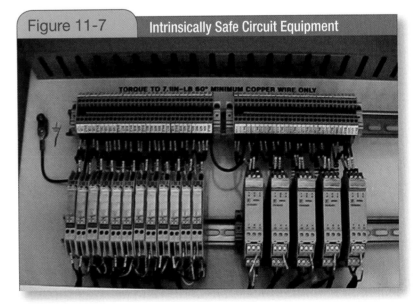

Figure 11-7 | Intrinsically Safe Circuit Equipment

Figure 11-7. The intrinsically safe circuit equipment limits the electrical energy on the conductors to the level that a spark cannot occur.

Class I, Zone 2
- Locations in which ignitable concentrations are not likely and could only occur briefly
- Locations in which volatiles are handled, processed, or used in closed containers
- Locations in which positive ventilation prevents accumulation of gases/vapors
- Areas adjacent to Class I, Zone 1 locations

Article 506 Zone 20, 21, and 22 Locations for Combustible Dusts or Ignitable Fibers/Flyings

506.5(B) Zone 20, Zone 21, and Zone 22 Locations

(1) Zone 20
- Locations in which ignitable concentrations of combustible dust or ignitable fibers or flyings are present continuously
- Locations in which ignitable concentrations of combustible dust or ignitable fibers or flyings are present for long periods of time

(2) Zone 21
- Locations in which ignitable concentrations of combustible dust or ignitable fibers or flyings exist under normal operation.
- Locations in which ignitable concentrations of combustible dust or ignitable fibers or flyings may exist due to repair, maintenance, or leaks
- Locations in which ignitable concentrations of combustible dust or ignitable fibers or flyings exist due to processes, breakdown, or faulty equipment
- Areas adjacent to Zone 20 locations

(3) Zone 22
- Locations in which ignitable concentrations of combustible dust or ignitable fibers or flyings are not likely to exist and could only occur briefly
- Locations in which combustible dust or ignitable fibers or flyings are handled, processed, or used in closed containers
- Areas adjacent to Zone 21 locations

Article 510 Hazardous (Classified) Locations - Specific

Specific Class I, II, and III locations - **Article 510** sets the stage for the application of **Article 511** through **Article 517**. Note that **Article 510** references **Article 517** for health care facilities because of the explosive gases or vapors which may be used in inhalation anesthetizing locations. **Article 517** is not grouped in this textbook with the specific Class I, II, and III locations due to the many other requirements which make **Article 517** special.

Article 510 clearly states that the provisions of **Article 500** through **Article 504** apply to the specific, listed locations unless modified or supplemented by **Article 511** through **Article 517**.

Article 510 clearly points out that the listed "specific hazardous location" articles modify or supplement **Article 500** through **Article 504**. These articles cover occupancies which are or may be hazardous, and the rules contained within modify or supplement **Article 500** through **Article 504**.

Article 511 Commercial Garages, Repair and Storage

Article 511 regulates service and repair operations for cars, buses, trucks, and tractors where volatile, flammable liquids, or gases are used for fuel or power. **See Figure 11-8.**

Article 513 Aircraft Hangars

Article 513 regulates buildings or structures containing aircraft and volatile flammable liquids.

Article 514 Motor Fuel Dispensing Facilities

Article 514 covers gas stations, marine/motor fuel dispensing facilities, and the dispensing of motor fuel indoors and outdoors.

Figure 11-8 | Auto Repair

Figure 11-8. Commercial garages are hazardous locations due to the presence of volatile liquids and/or gases as a fuel for vehicles.

Figure 11-9 | Health Care

Figure 11-9. Chapter 5 of the NEC details the special installation requirements for facilities such as hospitals (Article 517).

Article 515 Bulk Storage Plants

Article 515 covers all property where flammable liquids are received by tank vessel, pipelines, tank car, or tank vehicle and are stored or blended in bulk for the purpose of distributing such liquids by tank vessel, pipeline, tank car, tank vehicle, portable tank, or container.

Article 516 Spray Application, Dipping, and Coating Processes

Article 516 covers locations in which regular or frequent application of flammable or combustible liquids and powders occurs by spray operations.

Health Care Facilities

Article 517 Health Care Facilities

Health care facilities are special occupancies which require modification of the general rules in Chapters 1 through 4 of the *NEC*. These facilities occupy buildings or portions of buildings in which they provide medical, dental, psychiatric, nursing, obstetrical, or surgical care. **See Figure 11-9.** Health care facilities include, but are not limited to, hospitals, nursing homes, limited care facilities, clinics, medical and dental offices, and ambulatory care centers, whether permanent or movable.

A total of 40 definitions are included in **Section 517.2** to deal with the special requirements of a health care facility. Additionally, many various types of locations are covered by **Article 517**. The definition of health care facilities in **Section 517.2** outlines locations covered by this article. **Article 517** is divided into seven parts:

Article 517 Health Care Facilities

I. **General**

II. **Wiring and Protection**

III. **Essential Electrical System**

IV. **Inhalation Anesthetizing Locations**

V. **X-Ray Installations**

VI. **Communications, Signaling Systems, Data Systems, Fire Alarm Systems, and Systems Less Than 120 Volts, Nominal**

VII. **Isolated Power Systems**

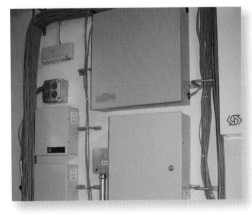

Part VI covers requirements for Limited Energy systems in Health Care facilities (such as Security and access control systems).

Assembly Occupancies

Article 518 Assembly Occupancies

The *NEC* provides special consideration of electrical installations to all locations where large numbers of people are intended to assemble. **See Figure 11-10. Article 518 Assembly Occupancies** contains modifications and supplemental requirements for all assembly locations. Per **Section 518.1**, this article is intended to cover all buildings or portions of buildings or structures intended for groups of 100 or more persons assembled for meetings, worship, entertainment, eating, drinking, amusement, awaiting transportation, or any other similar purpose. Examples of assembly occupancies include, but are not limited to, the following:

Armories	Assembly Halls	Auditoriums
Bowling Lanes	Club Rooms	Conference Rooms
Courtrooms	Dance Halls	Dining/ Drinking Facilities
Exhibition Halls	Gymnasiums	Mortuary Chapels
Multipurpose Rooms	Museums	Places Of Awaiting Transportation
Places Of Religious Worship	Pool Rooms	Restaurants
Skating Rinks		

Figure 11-10. Worship centers are one of the most common examples of an assembly occupancy.

Figure 11-11. Movie Theatres are entertainment venues typically with more the 100 occupants, therefore falling under the special conditions of Chapter 5.

Entertainment Venues

Entertainment venues have many special electrical installation requirements. For example, theatres and motion picture venues have special needs for lighting and equipment and are designed to be attended by 100 or more persons, making them assembly occupancies. **See Figure 11-11.** These locations are also intended to be occupied in low light level situations. Circuses, carnivals, and fairs also have very special needs due to their outdoor venue, in all types of weather, and large numbers of people in attendance.

Article 520 Theatres, Audience Areas of Motion Picture and Television Studios, Performance Areas, and Similar Locations

Article 522 Control Systems for Permanent Amusement Attractions

Article 525 Carnivals, Circuses, Fairs, and Similar Events

Article 530 Motion Picture and Television Studios and Similar Locations

Article 540 Motion Picture Projection Rooms

Agricultural Buildings for Poultry, Livestock, and Fish

Article 547 Agricultural Buildings

Agricultural buildings containing poultry, livestock, or fish present very special needs for electrical installations. One example is the requirement of **Section 547.10** for equipotential planes, a bonding requirement necessary to protect livestock from electric shock.

Special requirements for agricultural buildings, as well as buildings used for poultry, livestock, and fish confinement systems, are necessary where excessive dust and dust mixed with water might accumulate on electrical equipment.

Manufactured Buildings, Dwellings, and Recreational Structures

Special electrical construction and installation requirements exist when buildings or structures are manufactured and then moved into place for use by persons for occupancy including storage, places of employment, dwelling units, and recreational uses.

Article 545 Manufactured Buildings
Article 550 Mobile Homes, Manufactured Homes, and Mobile Home Parks
Article 551 Recreational Vehicles and Recreational Vehicle Parks
Article 552 Park Trailers

Structures on or Adjacent to Bodies of Water

Electrical installations for buildings and structures designed for use floating on, above, or adjacent to bodies of water require special electrical installation considerations. **See Figure 11-12. Article 553** and **Article 555** address the special needs of these occupancies.

Article 553 Floating Buildings
Article 555 Marinas and Boatyards

The definition in **553.2** describes a floating building as a building that floats on water and is moored in a permanent location and has permanent wiring to an electrical supply system.

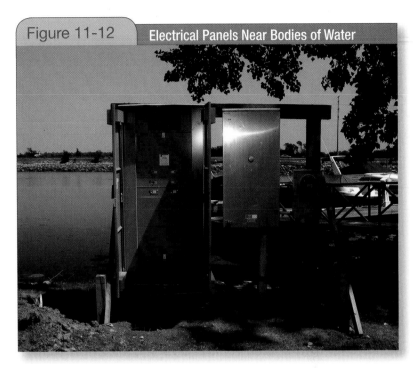

Figure 11-12 | **Electrical Panels Near Bodies of Water**

Figure 11-12. Articles 553 and 555 detail the requirements for electrical installations near bodies of water.

Temporary Installations

Article 590 Temporary Installations

Temporary power requires special installation considerations when used for construction, holiday decorative lighting, emergencies, or tests. **See Figure 11-13.**

Figure 11-13 | **Temporary Power**

Figure 11-13. Article 590 allows for temporary power installations for construction and other considerations.

Key provisions of **Article 590 Temporary Installations** are located in the following sections. **See Figure 11-14.**

590.3 Time Constraints

(A) During Period of Construction. Temporary wiring is permitted for the length of the project.

(B) 90 Days. Holiday and similar lighting is not permitted to exceed 90 days.

(C) Emergencies and Tests. Temporary installations are permitted for emergencies and tests.

(D) Removal. Temporary installations must be removed immediately after use.

590.6 Ground-Fault Protection for Personnel

(A) Receptacle Outlets. All 125-volt, single-phase, 15-, 20-, and 30-ampere receptacle outlets that are in use by personnel must be GFCI protected. When a receptacle is part of the permanent wiring of a building or structure and is intended for use by persons, GFCI protection listed as "portable" must be provided.

NEC CHAPTER 6, "SPECIAL EQUIPMENT"

The *Codeology* title for Chapter 6 is "Special Equipment," whereas Chapter 4 of the *NEC* is titled "Equipment for General Use" and covers the basics. *NEC* Chapter 6 covers special equipment and modifies or supplements the first four chapters through 25 articles to meet the special needs of the equipment covered in this chapter.

Figure 11-14 **Temporary Lighting Installation**

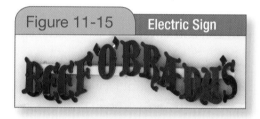

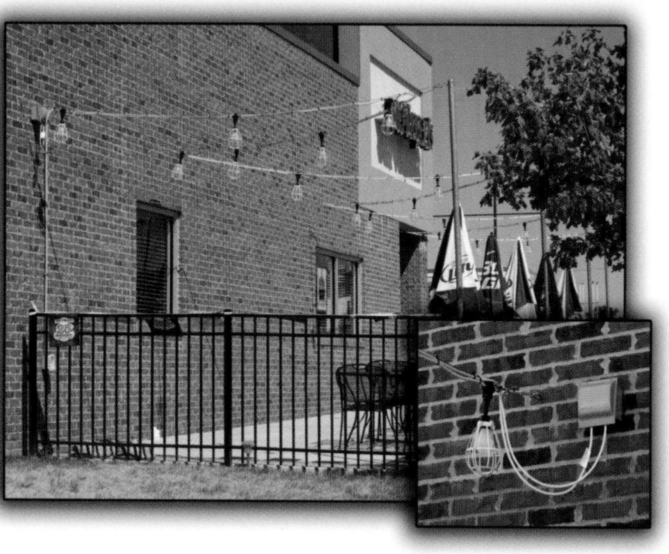

Figure 11-14. Article 590 allows for temporary lighting installations.

Article 600 Electric Signs and Outline Lighting

Article 600 covers all electric sign and outline lighting installations; including neon tubing. **See Figure 11-15.**

Figure 11-15 **Electric Sign**

Figure 11-15. Electric signs are special equipment and are covered in Chapter 6 of the NEC.

Article 604 Manufactured Wiring Systems

Article 604 covers field-installed wiring using manufactured subassemblies.

Article 605 Office Furnishings (Consisting of Lighting Accessories and Wired Partitions)

Article 605 covers wiring and lighting equipment on relocatable wired partitions.

Article 610 Cranes and Hoists

All electrical equipment and wiring used with cranes, monorail hoists, hoists, and all runways are addressed by **Article 610.**

Article 620 Elevators, Dumbwaiters, Escalators, Moving Walks, Platform Lifts, and Stairway Chairlifts

Article 620 covers the installation of electrical equipment and wiring for elevators, dumbwaiters, escalators, moving walks, platform lifts, and stairway chairlifts.

Article 625 Electric Vehicle Charging System

Conductors and equipment external to an electric vehicle related to charging the electric vehicle are covered by **Article 625**.

Article 626 Electrified Truck Parking Spaces

Article 626 covers electrical conductors, devices and equipment which connect trucks (tractor-trailer rigs) and associated refrigeration units to a source of electrical energy.

Article 630 Electric Welders

Article 630 covers the equipment and installation necessary for electric arc welding, resistance welding, plasma cutting, and other welding/cutting methods. **See Figure 11-16.**

Article 640 Audio Signal Processing, Amplification, and Reproduction Equipment

Article 640 covers equipment and wiring for distribution of sound, public address, speech input systems, temporary audio system installations, electronic organs or other electronic musical instruments, and audio signal generation, recording, processing, amplification, and reproduction.

Article 645 Information Technology Equipment

Article 645 covers information technology equipment and systems in an information technology room only where the conditions of **Section 645.4** are met.

Article 647 Sensitive Electronic Equipment

Installation and wiring of separately derived systems operating at 120 volts line to line and 60 volts to ground are covered by **Article 647**.

Article 650 Pipe Organs

Article 650 covers electrical circuits and parts of pipe organs which provide control of keyboards and sound.

Article 660 X-Ray Equipment

All X-ray equipment operating at any frequency or voltage for industrial or other nonmedical or nondental use is covered by **Article 660**.

Article 665 Induction and Dielectric Heating Equipment

Industrial and scientific applications of dielectric heating, induction heating, induction melting, and induction welding are covered by **Article 665**.

Article 668 Electrolytic Cells

Article 668 covers the use of electrolytic cells for the production of aluminum, cadmium, chlorine, copper, fluorine, hydrogen peroxide, magnesium, sodium, sodium chlorate, and zinc.

Article 669 Electroplating

The installation of electrical components and equipment for electroplating, anodizing, electropolishing, and electrostripping is covered by **Article 669**.

Figure 11-16   **Electric Welder**

For additional information, visit qr.njatcdb.org
Item #1075

Figure 11-16. Electric welders are covered in Article 630 of the NEC because of their operation on a duty cycle.

Article 670 Industrial Machinery

Article 670 covers overcurrent protection for industrial machinery, as well as the size of its supply conductors and the requirements for its nameplate data.

Article 675 Electrically Driven or Controlled Irrigation Machines

Electrically driven or controlled irrigation machines and branch circuits and controllers involved are covered by Article 675.

Article 680 Swimming Pools, Fountains, and Similar Installations

Article 680 covers the construction and installation of electrical wiring for and equipment in or adjacent to all swimming, wading, therapeutic, and decorative pools, fountains, hot tubs, spas, and hydro massage bathtubs (whether permanently installed or storable), and to metallic auxiliary equipment such as pumps, filters, and similar equipment.

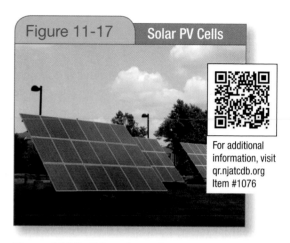

Figure 11-17   Solar PV Cells

For additional information, visit qr.njatcdb.org Item #1076

Figure 11-17. Solar photovoltaic systems are special equipment addressed in Chapter 6.

Article 682 Natural and Artificially Made Bodies of Water

Article 682 applies to all electrical installations made in or adjacent to natural or artificially made bodies of water not covered by other articles in the *NEC*.

Article 685 Integrated Electrical Systems

Article 685 covers integrated electrical systems when an orderly shutdown is necessary to ensure safe operation.

Article 690 Solar Photovoltaic (PV) Systems

Solar photovoltaic electrical energy systems, including the array circuits, inverters, and controllers are covered by Article 690. **See Figure 11-17.**

Article 692 Fuel Cell Systems

Stand-alone or interactive fuel cell power systems are covered by **Article 692.** See Figure 11-18.

Article 694 Small Wind Electric Systems

Small wind (turbine) electric systems up to 100 kW (inclusive) are covered by **Article 694.** Components such as generators, alternators, inverters, batteries, and controllers are part of this article. Several definitions specific to wind systems are introduced in **Section 694.2 Definitions.** This is an emerging market; therefore, the *Code* user should become familiar with the requirements within.

Figure 11-18   Residential Fuel Cell

Figure 11-18. Fuel cell systems are special equipment addressed in Chapter 6 (Article 692).

Article 695 Fire Pumps

Article 695 covers the installation of power sources, interconnecting circuits, and switching and control equipment for fire pumps. Note that this article does not apply to pressure maintenance or jockey pumps. **See Figure 11-19.**

NEC CHAPTER 7, "SPECIAL CONDITIONS"

The *Codeology* title for Chapter 7 is "Special Conditions" and covers conditions required to meet the special needs of different types of occupancies and equipment. For example, when an alternate power source is required for emergency or legally required standby, it must be installed according to Chapters 1 through 4 and modified and/or supplemented in accordance with the requirements of Chapter 7 Special Conditions.

Article 700 Emergency Systems

Article 700 applies to the installation, operation, and maintenance of emergency systems consisting of circuits and equipment intended to supply, distribute, and control electricity for illumination, power, or both. **See Figure 11-20.**

Article 701 Legally Required Standby Systems

Article 701 applies to the installation, operation, and maintenance of legally required standby systems consisting of circuits and equipment intended to supply, distribute, and control electricity for illumination, power, or both.

Article 702 Optional Standby Systems

Article 702 covers the installation and operation of optional standby systems.

Article 705 Interconnected Electric Power Production Sources

Electric power production sources which operate in parallel with a primary source are covered by **Article 705.** Examples of a primary source include a utility supply or an on-site electric power source.

Article 708 Critical Operations Power Systems (COPS)

Article 708 applies to designated critical operations areas when required by municipal, state, federal, or other codes. The installation, operation, monitoring,

Figure 11-19 **Fire Pump and Motor**

Figure 11-19. Fire pumps are very important to the safety of building occupants. Electrical installation requirements are covered in Article 695.

Figure 11-20 **Emergency System**

Figure 11-20. Emergency systems use a standby generator as the emergency source in many installations and represent special conditions addressed in Chapter 7.

control, and maintenance of premises wiring systems supplying electricity to critical operations areas are addressed to ensure their survivability in the event of naturally-occurring hazards and human-caused events.

Article 720 Circuits and Equipment Operating at Less Than 50 Volts

DC or AC installations operating at less than 50 volts are covered by **Article 720.**

Article 725 Class 1, Class 2, and Class 3 Remote-Control, Signaling, and Power-Limited Circuits

Remote control, signaling, and power limited circuits not part of a device or appliance are covered by **Article 725.**

Article 727 Instrumentation Tray Cable: Type ITC

Article 727 covers the construction specifications, use, and installation of instrumentation tray cables limited to applications at 150 volts or less and 5 amps or less.

Article 760 Fire Alarm Systems

All wiring, equipment, and circuits controlled by a fire alarm system are covered by **Article 760. See Figure 11-21.**

Article 770 Optical Fiber Cables and Raceways

The installation of all fiber cables and their raceways are covered by **Article 770. See Figure 11-22.**

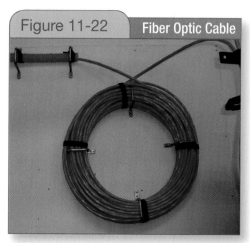

Figure 11-22. When fiber optic cable and raceways are installed, they represent a special condition addressed in Chapter 7.

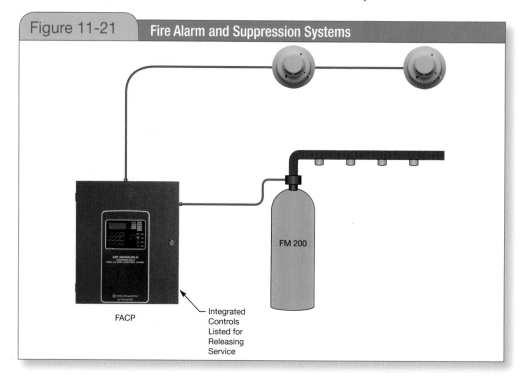

Figure 11-21. Fire alarm systems represent special conditions and are addressed in Chapter 7. Further requirements are also listed in the National Fire Alarm and Signaling Code, NFPA-72.

Summary

C hapters 5, 6, and 7 are known as the "Special Chapters" due to the fact that they contain modifications, supplemental information, and rules which complement, or complete the basic foundation of Chapters 1 through 4. These special chapters contain 63 articles which the Code user must become familiar with for proper application. When applying the rules of the *NEC* to any electrical installation, it is imperative that the user recognizes and refers to the special chapters whenever special occupancies, special equipment, or special conditions are part of an installation.

Review Questions

1. Do Chapters 5, 6, and 7 of the *NEC* apply to all electrical installations covered by its requirements?

2. Chapter 6 of the *NEC* is divided into how many articles?

3. The scope of Chapter 5 of the *NEC* is dedicated to special __?__.

4. Name the part and article which would address the installation of a temporary audio installation.

5. Chapter 7 of the *NEC* addresses special conditions. Name the part and article in Chapter 7 which would apply when sizing overcurrent protection in a legally required standby system.

6. Do the definitions in Section 500.2 apply to Article 501?

7. Which five articles in Chapter 5 address entertainment venues?

8. Article 690 is divided into __?__ parts.

9. Does Part III of Article 725 apply to Class 1 circuits?

10. Which special article covers requirements for generation of power with fuel cells?

11. What section of Article 694 covers charge control for wind systems?

To access Practice Problems, visit qr.njatcdb.org. Item #1082 Click on Inside Blended Learning.

Chapter 8 of the *NEC*, "COMMUNICATIONS SYSTEMS"

The *Codeology* title for Chapter 8 of the *NEC* is "Communications." Many would say that this chapter is an island, standing apart from the rest of the *Code*.

90.3 Code Arrangement

Chapter 8 covers communications systems and is not subject to the requirements of Chapters 1 through 7 except where the requirements are specifically referenced in Chapter 8.

Five articles comprise Chapter 8, the 800-series. These five articles provide the general requirements and information for all installations of communications systems.

Objectives

» Associate the *Codeology* title for *NEC* Chapter 8 as "Communications."

» Identify the type of information and requirements for communications systems covered in Chapter 8.

» Recognize Chapter 8 as the 800-series.

» Recognize, recall, and become familiar with articles contained in Chapter 8.

Chapter **12**

Table of Contents

NEC CHAPTER 8, "COMMUNICATIONS SYSTEMS"

Chapters 1 through 7 of the *NEC* do not apply to any of the five articles of Chapter 8 unless specific reference is made within these articles to another area of the *NEC*. For example, all five articles of Chapter 8 specifically reference **Article 100 Definitions** and each recognizes that all of the definitions in **Article 100** are applicable within that article. Four of the articles (800, 820, 830, and 840) include additional definitions which are applicable only within that article:

- **Article 800 - Section 800.2** recognizes all of Article 100 and adds eleven definitions which apply only within Article 800.
- **Article 810 - Section 810.2** recognizes all of Article 100 and adds no additional definitions.
- **Article 820 - Section 820.2** recognizes all of Article 100 and adds five definitions which apply only within Article 820.
- **Article 830 - Section 830.2** recognizes all of Article 100 and adds seven definitions which apply only within Article 830.
- **Article 840 - Section 840.2** recognizes all of Article 100 and adds four definitions which apply only within Article 840.

Each of the five articles of Chapter 8 dedicates a section for referencing other articles in the third section of the article (for example, **Section 800.3**). Among the articles recognized are the requirements in Chapter 5 where hazardous locations are encountered and **300.22(C)** for permitted wiring methods in space used for environmental air. These examples are not all inclusive; multiple other references to requirements in Chapters 1 through 7 appear in the five articles of Chapter 8.

While the rest of the *NEC* does not apply to Chapter 8 unless specifically referenced, all five articles of Chapter 8 must be enforced once the *Code* is adopted. The *NEC* covers signaling and communications conductors, equipment, and raceways and is intended to give governmental agencies legal jurisdiction over electrical installations including signaling and communications systems, per **Section 90.4**. Articles of Chapter 8 cover specific types, methods, conductors, and equipment for communications systems. **See Figure 12-1.**

ARTICLE 800 COMMUNICATIONS CIRCUITS

Article 800 is divided into six parts to cover the following installations: **See Figure 12-2.**

- Telephone installations
- Telegraph systems (except radio telegraph)
- Outside wiring for fire alarm and burglar alarm systems
- Telephone systems: equipment, installation, and maintenance

The six parts of **Article 800**:

I. **General**
II. **Wires and Cables Outside and Entering Buildings**
III. **Protection**
IV. **Grounding Methods**
V. **Installation Methods Within Buildings**
VI. **Listing Requirements**

Figure 12-1	Layout of *NEC* Chapter 8		
NEC® Title:	Communications Systems		
Codeology Title:	Communications		
Chapter Scope:	*Communications Systems* Only		
Article	**Article Title**		
800	Communications Circuits		
810	Radio and Television Equipment		
820	Community Antenna Television and Radio Distribution Systems		
830	Network-Powered Broadband Communications Systems		
840	Premises-Powered Broadband Communications Systems		

Figure 12-1. Chapter 8 contains five articles which, for the most part, stand alone from the rest of the chapters in the NEC.

800.2 Definitions list several additional definitions which are somewhat unique to the other chapters of the *NEC*. It is important for the Electrical Worker to understand these terms, for example:

- **Abandoned Communications Cable -** <u>A communications cable not terminated at both ends and not meant for future use</u>. This refers to an enormous problem which occurs often above the lay-in ceilings in office spaces. Because communications system technology is continuously improving, often new styles of cables are required to be installed. Unfortunately, the old, abandoned cables are not removed as required by **Section 800.25**. Over time, the space above the lay-in ceiling becomes quite full of abandoned cables, which poses problems with space and weight limits.

- **Communications Circuit Integrity (CI) Cable -** <u>Critical communications cables of emergency systems must provide continued operation</u>. As you can imagine, during an emergency situation, the communications cable cannot fail. The *NEC* details many requirements to ensure the high possibility of protection and operation for such cables.

- **Point of Entrance -** <u>The point where the communications cable enters the building through an outside wall</u>. The point of entrance is many times referred to as the demarcation point, the point where the system provider enters the building and the building owner distributes the operation.

- **Premises -** <u>The land and buildings that the communications system cables are installed</u>. Often times this is called the campus.

Figure 12-2. Chapter 8 covers the installation requirements for communications cables such as telephone systems.

Figure 12-2 | **Telephone M66 Blocks**

Fact

In Article 770 and in Chapter 8 of the 2011 *NEC*, the term "Grounding Electrode" has been replaced with either "Grounding Electrode Conductor" or "Bonding Conductor," depending upon the individual requirement.

Grounding

An important part of Chapter 8 which is often taken lightly by the Electrical Worker is grounding of communications systems. **Part IV** of Chapter 8 covers grounding and bonding methods required for communications systems. Not only does proper grounding reduce safety hazards, but most often ensures proper system performance. You have probably heard static noise on a telephone line before. This is typically the result of poor grounding in the system. If you apply that same idea to a communications cable transmitting data, you can imagine that the performance and accuracy of the data will suffer. Proper installation of the bonding conductor or grounding electrode conductor (GEC) is covered intensely throughout Chapter 8.

Raceways

Communications conductors can be installed in various raceways as described in a portion of **Part V** of Chapter 8. **800.100(A)-(L)** thoroughly list uses of various raceways throughout various environmental spaces.

Unlike power conductors, which can be grouped together in the same raceway, grouping power and communications cables in the same raceway <u>will</u> cause performance issues for the communications system. **800.133 Installation of Communications Wires, Cables, and Equipment** covers the separation of communications cables from power cables. As an installer, this section must be understood in detail.

Innerduct can be used to separate different communications cables within the same conduit or raceway. When a larger conduit is installed, it allows for additional cables of different systems to be added in the future. **See Figure 12-3.**

It is common to install communications conductors without the use of raceways above lay-in ceilings. The *NEC* still

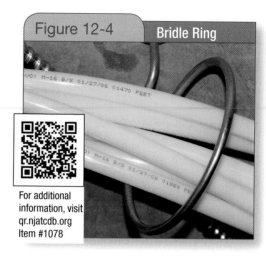

Figure 12-4   Bridle Ring

For additional information, visit qr.njatcdb.org Item #1078

Figure 12-4. It is common to install communications cables above a lay-in ceiling with bridle rings (or other cable management methods) instead of conduit.

requires a means of support for the conductors. Communications rings are often used since they do not damage the cable and are easy to install. **See Figure 12-4.** Please note, cables cannot be attached to the lay-in ceiling support wires; or, for that matter, even strapped to conduit. Strapping cables to a conduit will affect the ability for heat to escape, therefore reducing the ability for the power conductor to carry the proper current.

Listing Requirements

Part VI of Chapter 8 covers the required listing of the communications cable's outer jacket. Most communications systems installations do not require the cables to be installed in raceways, however, there are requirements in the *NEC* which cover installations and wiring methods for communications systems. **Section 800.154** describes the listing (rating) of the communications cable jacket in certain installations. For example, often times the space about the lay-in ceiling is known as a plenum space. A plenum space is where there is no supply or return ductwork for the HVAC. In most plenum situations, this space provides the return air path back to the HVAC equipment. Because of this situation, the jacket on the communications

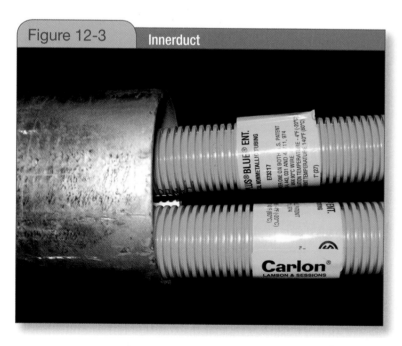

Figure 12-3   Innerduct

Figure 12-3. Installing innerduct in conduit allows different communications system's cables to be separated. This also allows for future cable installations.

cables must be of a higher grade to withstand the constant air movement and other factors involved. Though it is beyond the intent of this *Codeology* text to explain the differences between CMP, CMR, CMG/CM, CMX, and CMUC rated jackets, it would be advisable that the Electrical Worker become very familiar with **Part VI** of Chapter 8 before purchasing and installing communications cables outside of raceways in plenum ceilings. **See Figure 12-5**.

ARTICLE 810 RADIO AND TELEVISION EQUIPMENT

Article 810 consists of four parts which cover the following installations:

- Antenna systems for radio and television receiving equipment
- Amateur and citizen band radio transmitting and receiving equipment
- Transmitter safety
- Antennas and associated wiring and cabling, including the following types:
  See Figure 12-6:
  - Wire-strung type
  - Multi-element antennas
  - Vertical rod antennas
  - Dish antennas

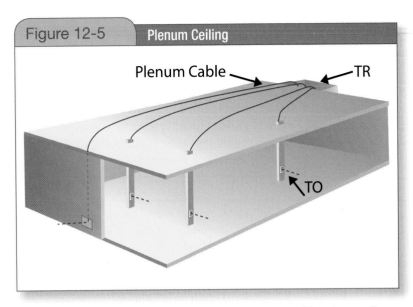

Figure 12-5 | **Plenum Ceiling**

Plenum Cable — TR

TO

Figure 12-5. It is common to install communications cables above a lay-in ceiling. But if not installed in a raceway, the cable shall be listed for plenum installations.

The four parts of **Article 810** are as follows:

I. **General**
II. **Receiving Equipment—Antenna Systems**
III. **Amateur and Citizen Band Transmitting and Receiving Stations—Antenna Systems**
IV. **Interior Installation—Transmitting Stations**

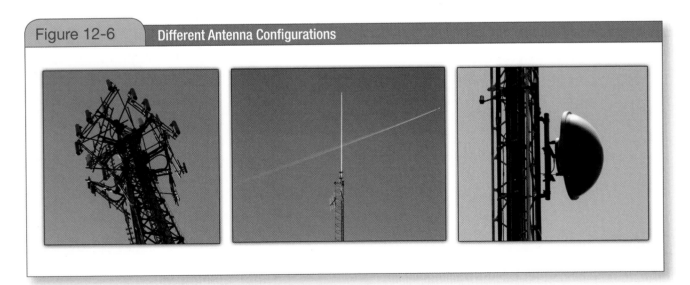

Figure 12-6 | **Different Antenna Configurations**

Figure 12-6. Article 810 covers antennas such as (left to right) multi-element, vertical rod, and dish antennas.

ARTICLE 820 COMMUNITY ANTENNA TELEVISION AND RADIO DISTRIBUTION SYSTEMS

All of the following six Parts of **Article 820** are dedicated to a single coverage: the cable distribution of radio frequency signals employed in community antenna television (CATV) systems. **See Figure 12-7.**

I. **General**
II. **Coaxial Cables Outside and Entering Buildings**
III. **Protection**
IV. **Grounding Methods**
V. **Installation Methods Within Buildings**
VI. **Listing Requirements**

ARTICLE 830 NETWORK-POWERED BROADBAND COMMUNICATIONS SYSTEMS

Article 830 is divided into six Parts to cover the following installations:

Network-powered broadband communications systems that provide any combination voice, audio, video, data, and interactive services through a network interface unit. **See Figure 12-8.**

The six parts of **Article 830** are as follows:

I. **General**
II. **Cables Outside and Entering Buildings**
III. **Protection**
IV. **Grounding Methods**
V. **Installation Methods Within Buildings**
VI. **Listing Requirements**

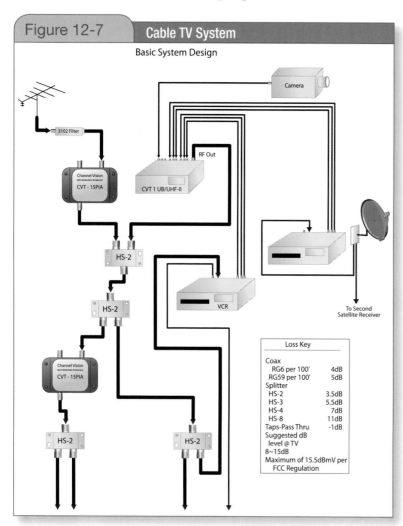

Figure 12-7. Article 820 lists the installation requirements for facility Cable TV systems.

Figure 12-8. Article 830 lists the installation requirements for facility data systems.

ARTICLE 840 PREMISES-POWERED BROADBAND COMMUNICATIONS SYSTEMS

Article 840 is divided into six Parts covering premises–powered optical fiber-based communications systems which provide any combination of voice, audio, video, data, and interactive services through an optical network terminal (ONT).

The six parts of **Article 840** are as follows:

I.   **General**
II.   **Cables Outside and Entering Buildings**
III.   **Protection**
IV.   **Grounding Methods**
V.   **Installation Methods Within Buildings**
VI.   **Listing Requirements**

> *Fact*
>
> Article 840 is a new article in the 2011 *NEC*. Previous editions did not specifically address premise-powered optical fiber based broadband communications systems.

An optical network terminal (ONT) is a device that converts an optical signal into component signals, including voice, audio, video, data, wireless, and interactive service electrical.

Article 770 Part IV contains the requirements for grounding and bonding of fiber optic cables with non–current-carrying metallic members.

Summary

Chapter 8, dedicated to communications systems, stands apart from the rest of the *NEC* in that the provisions of Chapters 1 through 7 apply to Chapter 8 only when a specific reference is made in a Chapter 8 article. Chapter 8 consists of the following four articles:

800	Communications Circuits
810	Radio and Television Equipment
820	Community Antenna Television and Radio Distribution Systems
830	Network-Powered Broadband Communications Systems
840	Premises-Powered Broadband Communications Systems

These five Chapter 8 articles cover specific methods, conductors, and equipment for communications systems. While other areas of the *NEC* may seem to be appropriately applied to a communications installation such as conduit or box fill, Chapter 8 is autonomous unless it specifically references any of the articles or sections in Chapters 1 through 7.

Review Questions

1. Do Chapters 1 through 7 of the *NEC* apply generally to all communications installations?

2. Chapter 8 is divided into how many articles?

3. The scope of Chapter 8 of the *NEC* is dedicated to circuits and equipment for all types of __?__ systems.

4. What part of which article in Chapter 8 would address the installation of cables inside a building for cable TV?

5. Chapter 8 of the *NEC* addresses communications systems. What part of which article would contain listing requirements for communications wire and cables?

6. When would a requirement located in Chapters 1 through 7 apply to an article located in Chapter 8?

7. What part of which article is referenced in 800.90(B)?

8. Article 830 is divided into how many parts?

9. Does Part II of Article 830 apply to grounding methods for network-powered broadband communications systems?

10. Which article in Chapter 8 covers requirements for amateur radio antenna systems?

11. Which article in Chapter 8 provides *Code* requirements for fiber optic communications systems?

To access Practice
Problems, visit
qr.njatcdb.org.
Item #1082
Click on Inside
Blended Learning.

Chapter 9 consists of twelve Tables and nine Informative Annexes. Tables are applicable only where referenced in the *NEC*, while Annexes are for informational use only and are not an enforceable part of the *Code*. These Tables and Informative Annexes supply the *Code* user with data to calculate conduit fill, ampacity, maximum raceway radius, voltage drops, and power source limitations. They also include product safety standards, types of construction classifications, functional performance tests (FPTs), SCADA applications, and many other examples.

Objectives

» Associate the *Codeology* title for *NEC* Chapter 9 as "Tables and Informative Annexes."

» Identify the specific type of information contained in the Tables and Informative Annexes of Chapter 9.

» Recognize that Tables in Chapter 9 apply only if referenced elsewhere in the *NEC*.

» Recognize that Informative Annexes are included in the *NEC* for informational purposes only.

» Recognize key words and clues for locating references to Chapter 9, the "Table and Informative Annex" chapter.

» Recognize, recall, and become familiar with Tables and Informative Annexes contained in Chapter 9.

Chapter 13

Table of Contents

TABLES

As required in **Section 90.3**, Chapter 9 contains tables which apply as referenced throughout the *NEC*. **See Figure 13-1.** As such, they are extremely valuable tools for the *Code* user. A basic understanding of the types of tables in Chapter 9 and where they are referenced for use is necessary for quick reference to the correct table.

Table 1 Percent of Cross Section of Conduit and Tubing for Conductors

Table 1 is the benchmark for all permissible combinations of conductors in conduit and is commonly referred to as the conduit fill table. **Table 1** is short and to the point. It lists only three types of conductor installations and permitted percentage of raceway fill. **Table 1** is referenced in the *NEC* wherever raceway fill requirements exist. For example, using rigid metal conduit, **Section 344.22** references the fill specified in **Chapter 9,** **Table 1.** The nine notes to **Table 1** explain the application of these rules in all applications. **See Figure 13-2.**

Table 2 Radius of Conduit and Tubing Bends

Table 2 provides a uniform minimum requirement for bends in conduit (to prevent damage to raceways) and reduction of the internal area of conductors. **Table 2** is referenced in the raceway articles. For example, using rigid metal conduit, **Section 344.24** references the bend requirements of **Chapter 9, Table 2.** Please note that in certain conduit installations, the job specifications may require a larger bend radius than what the *NEC* requires. This occurs because of the type of conductor which may be installed, such as fiber optic cables. In the case of fiber optic cables, larger radius conduit bends will prevent damage to the fiber and sustain the operation of light transmission to the fullest. **See Figure 13-3.**

Figure 13-1	NEC Chapter 9 Tables
NEC Chapter 9 Tables	
Table 1	Percent of Cross Section of Conduit and Tubing for Conductors
Table 2	Radius of Conduit and Tubing Bends
Table 4	Dimensions and Percent Area of Conduit and Tubing (Areas of Conduit or Tubing for the Combinations of Wires Permitted in Table 1, Chapter 9)
Table 5	Dimensions of Insulated Conductors and Fixture Wires
Table 5A	Compact Copper and Aluminum Building Wire Nominal Dimensions* and Areas
Table 8	Conductor Properties
Table 9	Alternating-Current Resistance and Reactance for 600-Volt Cables, 3-Phase, 60 Hz, 75°C (167°F) - Three Single Conductors in Conduit
Table 10	Conductor Stranding
Table 11(A)	Class 2 and Class 3 Alternating-Current Power Source Limitations
Table 11(B)	Class 2 and Class 3 Direct-Current Power Source Limitations
Table 12(A)	PLFA Alternating-Current Power Source Limitations
Table 12(B)	PLFA Direct-Current Power Source Limitations

Figure 13-1. The tables in Chapter 9 are used to calculate conductor fill along with conductor properties in calculating conductor size due to voltage drop.

Figure 13-2 Notes to *NEC* Table 1, Chapter 9

Notes to *NEC* Table 1, Chapter 9

Table 1   Percent of Cross Section of Conduit and Tubing for Conductors

Number of Conductors	All Conductor Types
1	53%
2	31%
Over 2	40%

Table 1 Notes (Installation Requirements)

1. Fixture wire conduit fill calculation references to Informative Annex C.

2. Table 1 requirements only apply to complete conduit installations and not applications of physical protection of exposed wiring.

3. Grounding and bonding conductors are to be included in conduit fill calculations.

4. 60% conduit fill is permissible within 24 inches or less conduit installations between boxes, etc.

5. Actual outside dimensions of conductors not listed in Chapter 9 shall be used in conduit fill calculations.

6. Tables 4, 5, and 5A shall be used for combinations of different size conductors in the conduit fill calculation.

7. The allowance to round-up to the next whole number in the case of conduit fill calculation with same size conductors

8. Requirements of Table 8 for bare conductors

9. Allowances to count multiple-conductor cables as one conductor in the conduit fill calculations

Figure 13-2. Jamming can occur when pulling multiple conductors in a conduit. Chapter 9, Table 1 lists the fill percentage of the conduit along with notes of installation requirements.

Figure 13-3   Minimum Conduit Radius

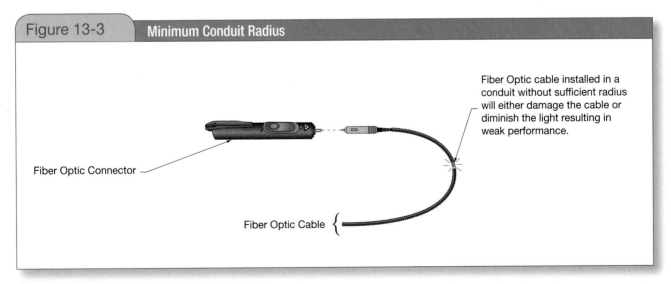

Fiber Optic cable installed in a conduit without sufficient radius will either damage the cable or diminish the light resulting in weak performance.

Fiber Optic Connector

Fiber Optic Cable

Figure 13-3. Not only does the NEC require a minimum radius in raceway installations, often times the job specifications increase the minimum radius.

Table 4 Dimensions and Percent Area of Conduit and Tubing (Areas of Conduit or Tubing for the Combinations of Wires Permitted in Table 1, Chapter 9)

Table 4 is extremely useful for the *Code* user when determining conduit fill. Information provided by this table includes total internal area and permissible fill area for several applications. **Table 4** is referenced in **Note 6** to **Table 1**, making **Table 4** applicable wherever **Table 1** is referenced in the *NEC*. **Table 4** is subdivided into twelve tables which list the various raceways from Chapter 3, such as **Article 358 Electrical Metallic Tubing: Type EMT, Article 362 Electrical Nonmetallic Tubing: Type ENT, Article 348 Flexible Metal Conduit: Type FMC**, etc. **See Figure 13-4.**

Table 5 Dimensions of Insulated Conductors and Fixture Wires

Table 5 provides the dimensions of insulated conductors and fixture wires needed to determine permissible conduit fill. **Table 5** is required to be used together with **Table 4** to determine permissible combinations of conduit fill. **Table 5** is referenced in **Note 6** to **Table 1**, making **Table 5** applicable wherever **Table 1** is referenced in the *NEC*.

Table 5A Compact Copper and Aluminum Building Wire Nominal Dimensions and Areas

Table 5A provides the dimensions of compact copper and aluminum building wire needed to determine permissible conduit fill. **Table 5** is required to be used together with **Table 4** to determine permissible combinations of conduit fill. **Table 5A** is referenced in **Note 6** to **Table 1**, making **Table 5A** applicable wherever **Table 1** is referenced in the *NEC*.

Table 8 Conductor Properties

Table 8 provides conductor properties for all conductor sizes from 18 AWG to 2,000 kcmil. Information provided includes circular mil area for all AWG sizes, stranding, diameter, and DC resistance. **Table 8** is referenced in **Note 8** to **Table 1** for determining area for bare conductors, making **Table 8** applicable wherever **Table 1** is referenced in the *NEC*.

Table 9 Alternating-Current Resistance and Reactance for 600-Volt Cables, 3-Phase, 60 Hz, 75°C (167°F)—Three Single Conductors in Conduit

Table 9 provides resistance and impedance values necessary for determining proper conductor application where voltage drop or other calculations are required. In long run installations of conductors, voltage drop is directly proportional to the length from the source to the load. A good example is the long conductor runs to parking lot lighting. Typically branch circuit conductor runs to parking lot light fixtures can be several hundred feet. Using **Table 9** allows the user to calculate the increased conductor size in order to minimize the voltage drop to the lighting poles.

Table 10 Conductor Stranding

Table 10 declares the number of strands for Class B and Class C copper conductors and Class B aluminum conductors. There are five classifications of conductor stranding: Class AA, A, B, C, and D. Class AA is only a few strands, or practically a solid conductor, whereas, Class D is the most finely stranded cable; such as a welding cable. The purpose of stranded cables in power applications is primarily

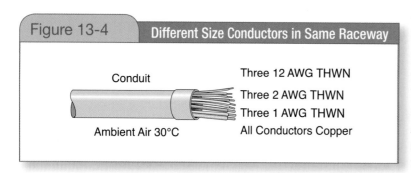

Figure 13-4   Different Size Conductors in Same Raceway

Conduit

Three 12 AWG THWN

Three 2 AWG THWN

Three 1 AWG THWN

Ambient Air 30°C   All Conductors Copper

Figure 13-4. Table 4 of Chapter 9 contains the information to calculate conduit size when filled with different size conductors.

for flexibility. The amount of strands in a cable typically determines the sizes of the strands; the more strands the larger the outside circular dimension of the conductor. There are various reasons why the *NEC* declares the maximum amount of strands. One reason is related to the termination of a high count stranded cable in a termination lug. If the strands are too small, they may not clamp down well in the termination.

Table 11(A) Class 2 and Class 3 Alternating-Current Power Source Limitations and Table 11(B) Class 2 and Class 3 Direct-Current Power Source Limitations

Table 11(A) and **Table 11(B)** provide the required power source limitations for listed Class 2 and Class 3 power sources.

Table 12(A) PLFA Alternating-Current Power Source Limitations and Table 12(B) PLFA Direct-Current Power Source Limitations

Table 12(A) and **Table 12(B)** provide the required power source limitations for listed PLFA power sources.

INFORMATIVE ANNEXES

As required in **Section 90.3**, Chapter 9 annexes are not part of the requirements of the *NEC*, but are included for informational purposes only. A basic understanding of the types of annexes in Chapter 9 and the information they contain is necessary for the *Code* user to access this valuable information. **See Figure 13-5.**

Informative Annex A Product Standards

Informative Annex A provides a list of UL, ANSI/ISA, and IEEE product safety standards used for the listing of products required to be listed in the *NEC*. These standards are very helpful for the installer for they not only provide vital information for safe installations, but also sound installation practices to ensure maximum performance of the equipment.

Figure 13-5	*NEC* Chapter 9 Informative Annexes
Informative Annex A	Product Safety Standards
Informative Annex B	Application Information for Ampacity Calculation
Informative Annex C	Conduit and Tubing Fill Tables for Conductors and Fixture Wires of the Same Size
Informative Annex D	Examples
Informative Annex E	Types of Construction
Informative Annex F	Availability and Reliability for Critical Operations Power Systems; and Development and Impletentation of Functional Performance Tests (FPTs) for Critical Operations Power Systems
Informative Annex G	Supervisory Control and Data Acquisition (SCADA)
Informative Annex H	Administration and Enforcement
Informative Annex I	Recommended Tightening Torque Tables from UL Standard 486A-B

Figure 13-5. The Informative Annexes in Chapter 9 are a great source of vital information for the installer; but are not part of the NEC requirements.

Informative Annex B Ampacities

Informative Annex B provides application information for many types of ampacity calculations, including those for conductors installed in electrical ducts.

Informative Annex C Tables

Informative Annex C is a useful aid to the user of the *Code* in determining conduit fill when all the conductors to be installed in a raceway are of the same size and type. **Informative Annex C** is informational only and is referenced in **Note 1** to **Table 1**. **See Figure 13-6.**

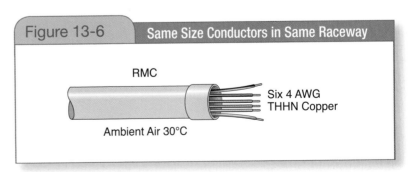

Figure 13-6. Informative Annex C of Chapter 9 contains the information to calculate conduit size when filled with conductors of the same size.

Informative Annex D Examples

Informative Annex D is provided to aid the *Code* user when making calculations required by the *NEC*. The requirements for calculations are illustrated in the form of examples to aid those making similar calculations. **Informative Annex D** is a "must read" for all electrical installers.

Informative Annex E Construction Types

Informative Annex E is provided to aid the *Code* user when determining any of the five types of building construction. Several tables are included in **Informative Annex E** which details fire resistance ratings, maximum number of stories (floors) per type of construction, and cross-references.

Informative Annex F Critical Operations Power Systems

The information in **Informative Annex F** is invaluable to the *Code* user when compliance with the commissioning requirements of **Article 708 Critical Operations Power Systems (COPS)** is necessary.

Informative Annex G SCADA

The information in **Informative Annex G** is useful when implementing a security control and data acquisition system which may be installed together with a critical operations power system described in **Article 708**.

Informative Annex H Administration

Informative Annex H is provided as a model set of administration and enforcement requirements which could be adopted by a governmental body along with the electrical installation requirements of the *NEC*. Chapter 14 of *Codeology* discusses test preparation. Typically, the test is administrated by the local authority having jurisdiction (AHJ). **Informative Annex H** is a sample of developing the basics of a governing body over electrical installations. This annex offers development for: inspections, fire investigations, review of construction documents, testing of electrical systems, and regulation of electrical installations.

Informative Annex I Tightening Torque Tables

In the world of electrical installations, torque applications are all around you. Today, torque application includes everything from switches and breakers to connectors and busbar. Manufacturers typically have recommended torque requirements for proper installation. **Informative Annex I** lists recommended torque levels as defined in UL Standard 486A-B. Keep in mind, the manufacturer's torque recommendations supersede all other listed standards. **See Figure 13-7.**

For additional information, visit qr.njatcdb.org Item #1077

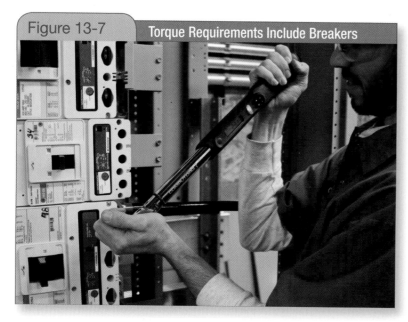

Figure 13-7 — Torque Requirements Include Breakers

Figure 13-7. The manufacturer's torque requirements must supersede the torque recommendations listed in Informative Annex I (UL Standards).

Summary

The twelve tables of Chapter 9 of the *NEC* apply only where referenced in the *Code*. These tables are necessary when applying many of the provisions of the *NEC*. Conduit fill calculations may include the use of several tables including the following:

Table 1   Percent of Cross Section of Conduit and Tubing for Conductors

Table 4   Dimensions and Percent Area of Conduit and Tubing (Areas of Conduit or Tubing for the Combinations of Wires Permitted in Table 1, Chapter 9)

Table 5   Dimensions of Insulated Conductors and Fixture Wires

Table 5A Compact Copper and Aluminum Building Wire Nominal Dimensions and Areas

Table 8   Conductor Properties

The *Code* Informative Annexes are intended only to aid the user of the *NEC*. These Informative Annexes offer useful information on product standards, conduit fill, types of construction, cross-reference tables, the application of **Article 708**, and administration and enforcement. Additionally, through the use of examples, **Informative Annex D** illustrates the calculations required in the *Code*.

Review Questions

1. **The tables located in Chapter 9 of the *National Electrical Code* apply __?__.**
   a. at all times
   b. only in Chapters 5, 6, and 7
   c. wherever they are useful
   d. as referenced in the *NEC*

2. **Which table is referenced in raceway articles for conduit fill?**
   a. Table 1
   b. Table 2
   c. Table 3
   d. Table 4

3. **When the *NEC* requires wiring methods, materials, and equipment to be listed, Informative Annex __?__ provides additional information on the product standards.**
   a. G
   b. D
   c. F
   d. A

4. **The Informative Annexes in Chapter 9 of the *NEC* are __?__.**
   a. not mandatory requirements
   b. mandatory requirements
   c. applicable as referenced
   d. used only for special equipment

5. **Table 10 covers conductor __?__.**
   a. ampacity
   b. properties
   c. insulations
   d. stranding

Test Preparation

This chapter is designed for those preparing to take an *NEC* proficiency exam. The key to success in any type of exam is to be fully prepared. However, a *Code* exam can be quite different from other exams which you may have taken in the past. Your ability to quickly and accurately find information in the *NEC* using the *Codeology* method and sound time management are the keys to success when taking an *NEC* proficiency exam. In addition, you must be able to recognize the various types of exams, questions, and testing methods before you can approach test day with any level of self-confidence.

Objectives

» Recognize the importance of the *Codeology* method when taking *NEC* exams.

» Identify the different types of exams.

» Identify the different types of exam questions.

» Recognize the importance of proper time management when taking an exam.

» Recognize the importance of developing a game plan before taking an exam.

Chapter **14**

Table of Contents

EXAMS BASED UPON THE *NEC*

For Electrical Contractor, Journeyman or Master Electrician, and Inspector

In many states, electrical contractors, electrical inspectors, and electrical workers are required to be licensed. Preparing for an *NEC* exam to become an electrical contractor, inspector, or electrician requires many hours of study. Three prerequisites for those about to take an *NEC* exam are: preparation, preparation, and more preparation. A method to quickly and accurately find the needed information is necessary for success when taking an *NEC* exam. The *Codeology* method is the most valuable tool for use on test day. Time management is the second most valuable tool.

Article-specific courses are necessary to obtain proficiency in areas such as grounding and bonding along with mastering calculations for services and feeder sizing, box and conduit fill, conductor ampacity corrections, and motor installations. All *NEC* exams are separated into groups of questions, the majority of which can be answered without time-consuming calculations. For example, in an exam of 80 questions, 50 or more will not require time-consuming calculations.

These 50 questions are those which can be quickly and accurately answered using the *Codeology* method. These questions are often closed book. The remaining questions are typically open book and can be answered by doing the required calculations after locating the correct requirements in the *NEC*. Using the *Codeology* method for the calculation questions will be vital.

TYPES OF EXAMS

Preparing for the exam includes contacting the testing organization and requesting information such as dates, times, submissions required prior to the day of testing, fees, and what you are allowed to bring with you into the testing room. Of course, the *NEC* will be allowed for the test, but many test situations do not allow the **NEC Handbook**. **See Figure 14-1.** The **NEC Handbook** contains the entire electrical *Code* (**NFPA 70**) with additional informative information to give the reader a clear explanation of the *Code*. In some cases, a *Code* book which contains highlights and written notes are not allowed. Most likely, an electrical calculation calculator will not be allowed. Be sure to confirm what is and is not allowed in respect to electrical theory textbooks as well. It would be unfortunate to arrive at the testing center to discover you are not able to take the test because your *Code* book contained tabs.

Written Exams

Written exams are slowly becoming a thing of the past due to the advantages of taking the exam online proctored at a local licensing identity. Nevertheless, the written test which utilizes a scanned answer form, is still used in some geographical areas. A written exam usually consists of an exam sheet containing multiple choice questions and an answer scan sheet. The submitted answers are not collected from the exam sheet, but from the answer scan sheet.

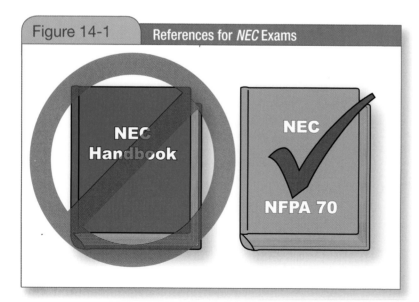

Figure 14-1	References for *NEC* Exams

Figure 14-1. Most NEC exams do not allow the NEC Handbook as a reference in the test. Often, notes written in the margins of the NEC are also not allowed.

Most applicants will use the exam sheet to perform their calculations and even mark the answer for reference. But, the answers must be marked on the answer scan sheet by filling in one of the four possible answer circles for each question. Once the testing session is over, the test administrator will deliver the answer scan sheets to the testing organization for scanning and returning of grades. **See Figure 14-2.**

When taking a written exam, it is vital that an applicant place each answer in its proper location. For example, if an applicant accidentally skips Question No. 2 on the exam sheet and fills in the answer for Question No. 3 in the space intended for Question No. 2, the exam would be completed with answers one question off on the answer scan sheet. This would be a real waste of the applicant's time.

Computer Exams

Computer-based exams are easily navigated for the computer user who possesses basic computer skills. On the other hand, applicants without basic computer skills should seek assistance and practice using a PC before test day. Most providers of *NEC* exams will have sample questions available on their web sites. All applicants, with or without basic computer

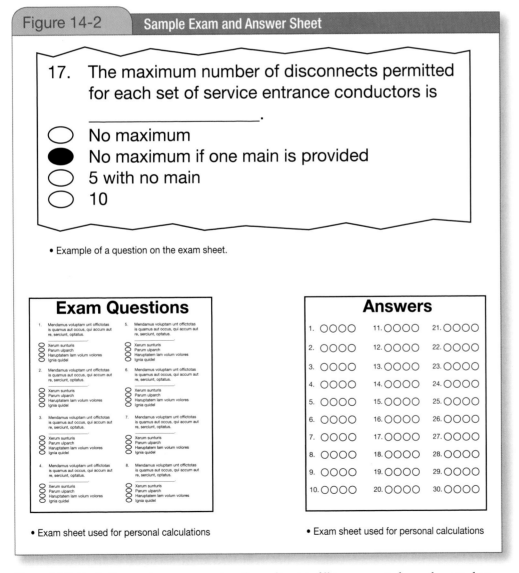

Figure 14-2. Typically, a written exam requires the applicant to fill in answer circles on the scan sheet.

skills, should take advantage of the opportunity to become familiar with the computer testing format. Additionally, in most cases, the exam administrator will offer a short tutorial on the exam format before the exam. One of the advantages of computer exams is receiving your performance grade shortly after the exam session. Keep in mind, you will still need to bring paper to perform the calculations before entering the answer on the computer screen. **See Figure 14-3.**

READING EACH QUESTION

Applicants should read each multiple choice question and all of the answers carefully before applying the *Codeology* method. It is essential that the question and all of the answers be read because a key word or clue may be in one of the possible answers. If the answer is not apparent, then consider highlighting the question to be solved later. By reading each question and all of the answers, you will subconsciously retain those questions. As you move through the exam, other questions and/or answers will jog your memory to solve a skipped question.

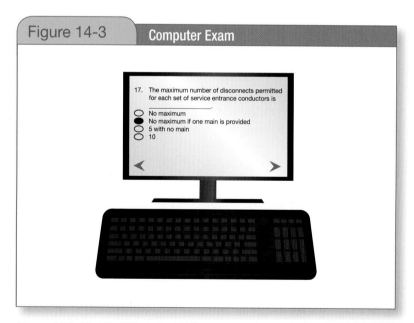

Figure 14-3   Computer Exam

Figure 14-3. The computer exam is the most popular method of taking the NEC exam today.

TYPES OF QUESTIONS

Code questions can be broken down into two basic solving categories: those which can be solved quickly and may involve a simple calculation, and those which require time-consuming calculations. Understanding this concept and applying the proper time management techniques throughout the exam will increase your chances of success. Time management is discussed in the next section.

General Knowledge Questions

General knowledge questions are included in all *NEC* exams. These questions are not designed to determine *NEC* proficiency, but to determine general knowledge of the electrical trade and electrical theory. General knowledge questions are often part of the closed book portion of the exam.

General questions often include the following material:

• Ohm's Law
• Power formula
• AC formulas
• Inductance and capacitance
• Conversions (for example, horsepower to watts)
• Voltage drop
• Local codes

Multiple Choice Questions

Typically, four answers follow each question in an *NEC* exam. When you know the answer, choose the correct option and move on to the next question. All questions must be answered by the end of the exam, unless there is no penalty for unanswered questions. For multiple-choice questions which are difficult to answer, try to eliminate one or more of the options. When more than one answer seems to be correct and "all of the above" is an option, it is probably the correct answer. If you are able to eliminate two of the four possibilities and you are forced to guess, the odds of getting a correct answer are increased. **See Figure 14-4.**

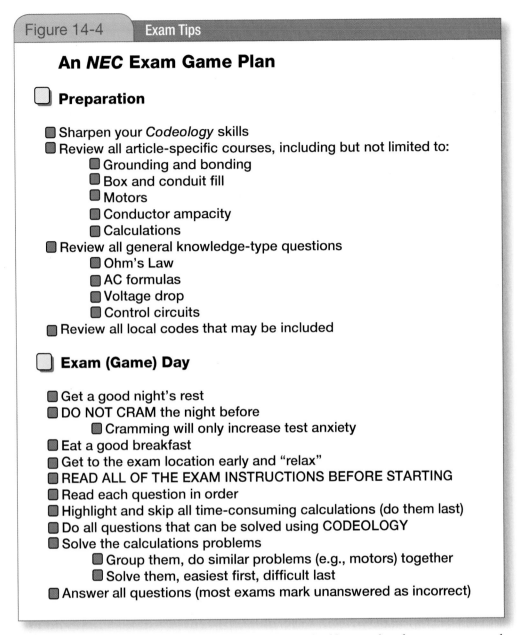

Figure 14-4   Exam Tips

An *NEC* Exam Game Plan

☐ **Preparation**

- ☐ Sharpen your *Codeology* skills
- ☐ Review all article-specific courses, including but not limited to:
    - ☐ Grounding and bonding
    - ☐ Box and conduit fill
    - ☐ Motors
    - ☐ Conductor ampacity
    - ☐ Calculations
- ☐ Review all general knowledge-type questions
    - ☐ Ohm's Law
    - ☐ AC formulas
    - ☐ Voltage drop
    - ☐ Control circuits
- ☐ Review all local codes that may be included

☐ **Exam (Game) Day**

- ☐ Get a good night's rest
- ☐ DO NOT CRAM the night before
    - ☐ Cramming will only increase test anxiety
- ☐ Eat a good breakfast
- ☐ Get to the exam location early and "relax"
- ☐ READ ALL OF THE EXAM INSTRUCTIONS BEFORE STARTING
- ☐ Read each question in order
- ☐ Highlight and skip all time-consuming calculations (do them last)
- ☐ Do all questions that can be solved using CODEOLOGY
- ☐ Solve the calculations problems
    - ☐ Group them, do similar problems (e.g., motors) together
    - ☐ Solve them, easiest first, difficult last
- ☐ Answer all questions (most exams mark unanswered as incorrect)

Figure 14-4. To be successful on the NEC exam, the applicant should review the information over several weeks instead of cramming a few days prior.

TIME MANAGEMENT

When taking an *NEC* exam, a given number of questions must be answered within a specified amount of time. For example, an 80 question exam could be given in a 4-hour period. Proper time management is essential when taking any exam. Additionally, most exams assign the same weight, or value, to each question. For example, in an 80 question, 4-hour exam, each question would be worth 1.25 points with only 3 minutes to solve. Time-consuming calculations may take 10 minutes or more while others can be solved in 1 minute or less using the *Codeology* method.

Thus, when taking an *NEC* exam, time-consuming calculations should always be skipped and solved last, regardless of how comfortable the applicant may feel in solving the problem. Interviews of applicants after taking an *NEC* exam typically reveal the two following scenarios:

Scenario 1

Marty is taking a 4-hour, 80-question electrical contractor exam. He solves each question in order. The front of the test is loaded with calculation-type questions. Marty is hung up on a few questions and his calculations do not match any of the answers. He checks his watch and realizes that 2 hours have passed and he is only on Question No. 18, therefore leaving only two hours to complete the remaining 62 questions. Panic sets in and Marty starts getting nervous. His self-confidence has dissipated and he begins to believe that he will fail the exam. Marty's ability to calmly read and solve each question is severely compromised. He did not use his time wisely.

Scenario 2

Virginia is taking a 4-hour, 80-question electrical contractor exam. She reads each question in order, highlighting the time-consuming, calculation-type questions for completion later and uses *Codeology* to quickly solve the other questions. Virginia checks her watch and realizes that 2 hours have passed and she has completed 55 questions, therefore having two more hours to complete the remaining 25 questions. She is pleased with her progress. Virginia's self-confidence has grown and she feels sure she will pass the exam. Virginia's ability to calmly read and solve each question has increased because of her renewed self-confidence. Unlike Marty, she has used her time wisely.

Summary

Preparing for and taking any type of *Code* exam whether for an entry-level position or to become an electrical contractor can be an unnerving experience for many electrical workers. Without a game plan, time management skills, and a method to quickly solve basic questions, the chances of passing the exam are not high for the ill-prepared applicant.

Proficiency in the *NEC* requires more than *Codeology* and time management skills. For most, it also requires taking article- or topic-specific *Code* courses to become competent in calculations, grounding, and other areas. However, on test day, the two most important tools for the applicant are:

- A method to quickly find needed information (*Codeology*)
- Time management

Applicants must familiarize themselves with the type of upcoming test, either written or computer-based. Applicants must also understand the type of questions to be included on an exam, the value of each question, the number of questions, the time allowed, and whether unanswered questions will be marked as incorrect. Time-consuming calculations should be left for completion at the end of the exam. Whether the exam is written or on a computer, the applicant must contact the testing organization and request information such as dates, times, submissions required prior to the day of testing, fees, and what reference materials can be allowed in the testing room.

Proper preparation for an exam will result in a high level of self-confidence, allowing the applicant to be more relaxed. Skipping time-consuming problems early in the exam and using *Codeology* to solve all the others will also result in increased self-confidence. It will also allow ample time near the end of the exam to solve calculations calmly and carefully.

1. *Code* users taking a competency exam on the National Electrical Code must take which of the following steps to be completely prepared?
    a. Practice sound time management
    b. Prepare and study for all types of *Code* questions
    c. Be capable of quickly and accurately finding information
    d. All of the above

2. Using proper time management methods, the *Codeology* user will skip all the time-consuming __?__ problems and solve them after all other questions have been answered.
    a. Chapter 5
    b. calculation
    c. multiple-choice
    d. grounding

3. General knowledge-type questions on *NEC* competency exams may include questions involving __?__.
    a. Ohm's Law and AC formulas
    b. control circuits
    c. voltage drop
    d. all of the above

4. Lack of proper preparation such as understanding time management skills, the ability to quickly and accurately find information, and article-specific preparation will result in reduced __?__ and ultimately, failure on an exam.
    a. self-confidence
    b. ability to guess
    c. good luck
    d. speed reading

5. Which of the following is/are likely to not be permitted in the testing room?
    a. *NEC* Handbook
    b. Electrical calculation calculators
    c. *Code* book with notes written in the margins
    d. All of the above